THE MIEGUNYAH PRESS

THE GENERAL SERIES OF THE
MIEGUNYAH VOLUMES
WAS MADE POSSIBLE BY THE
MIEGUNYAH FUND
ESTABLISHED BY BEQUESTS
UNDER THE WILLS OF
SIR RUSSELL AND LADY GRIMWADE.

'MIEGUNYAH' WAS THE HOME OF
MAB AND RUSSELL GRIMWADE
FROM 1911 TO 1955.

AUSTRALIA'S
REMARKABLE TREES

AUSTRALIA'S
REMARKABLE TREES

Richard Allen and Kimbal Baker

THE
MIEGUNYAH
PRESS

To: Archie, Max and Zoe Allen and
Raffaella, Romy & Isabella Baker

And our patient wives
Emma Allen & Deborah Baker

THE MIEGUNYAH PRESS
An imprint of Melbourne University Publishing Limited
187 Grattan Street, Carlton, Victoria 3053, Australia
mup-info@unimelb.edu.au
www.mup.com.au

First published 2009
Reprinted 2009, 2010
Paperback edition published 2010
Text © Richard Allen, 2009
Photography © Kimbal Baker, 2009
Design and typography © Melbourne University Publishing Limited, 2009

Designed by Pfisterer + Freeman
Printed in China through The Australian Book Connection

National Library of Australia Cataloguing-in-Publication entry

Allen, Richard, 1963–
Australia's remarkable trees / Richard Allen; photography
by Kimbal Baker.
9780522857887 (pbk.)
Trees—Australia.
Historic trees—Australia.
Baker, Kimbal.
582.160994

The mighty Ponderosa Pine at Khancoban Station *at the top of the Murray Valley in New South Wales, the Snowy Mountains a startling backdrop*

THROWING A TREE

I

The two executioners stalk along over the knolls,
Bearing two axes with heavy heads shining and wide,
And a long limp two-handled saw toothed for cutting great boles,
And so they approach the proud tree that bears the death-mark on its side.

II

Jackets doffed they swing axes and chop away just above ground,
And the chips fly about and lie white on the moss and fallen leaves;
Till a broad deep gash in the bark is hewn all the way round,
And one of them tries to hook upward a rope, which at last he achieves.

III

The saw then begins, till the top of the tall giant shivers:
The shivers are seen to grow greater with each cut than before:
They edge out the saw, tug the rope; but the tree only quivers,
And kneeling and sawing again, they step back to try pulling once more.

IV

Then, lastly, the living mast sways, further sways: with a shout
Job and Ike rush aside. Reached the end of its long staying powers
The tree crashes downward: it shakes all its neighbours throughout,
And two hundred years' steady growth has been ended in less than two hours.

THOMAS HARDY

FOREWORD

WHEN RICHARD AND KIM APPROACHED ME FOR SUPPORT WITH their *Remarkable Trees* project, I was keen to help. Their photographs and stories bear witness to their enthusiasm and willingness to travel to faraway places to find the best examples of Australia's rich variety of trees.

Trees are a vital part of our lives and this book reminds us how precious they are, and how important they remain for our future generations.

The importance of trees became obvious to me when I was a young jackaroo in Western Australia's Pilbara region. For those who haven't been fortunate enough to visit my beautiful, isolated outback home of the Pilbara, it's a vast stretch of wild land covering more than half a million square kilometres of Western Australia's central north region—a land of beauty and serenity, and harsh and unforgiving landscapes.

It was the height of summer and my truck broke down. The heat radiated up from a blistered road that stretched away in both directions as far as I could see. This heat quickly turned my four-wheel drive into an oven.

My only salvation from the oppressive heat was a beautiful Australian gum tree and a little clump of acacia bushes about 20 metres away. After an unsuccessful tinker with the car's engine, I gathered up a near-empty water bag and settled down under the shade of my new best friends to wait for search and rescue—I knew there was no way anyone could arrive before nightfall.

Although the tree had seen better days and was slowly losing its leaves to the hot winds that sweep through the region, its branches shielded me from the harsh rays of the sun overhead.

I have had furniture and lived in houses built from Red Gum and Jarrah, had many a picnic under a shady tree, climbed Ghost Gums and Karri as a child, and benefited in thousands of other ways from the presence of trees. Trees are also crucial in maintaining human life, wildlife and the environment on our planet. However, the importance of our leafy and green friends was never more apparent to me than during that searingly hot day, in the height of an inland Pilbara summer, spent under the life-saving shade of a beautiful local tree.

Richard's stories and Kim's evocative photographs will—I hope—encourage Australians to give our ancient and remarkable trees the respect and recognition they rightly deserve.

Andrew Forrest

CONTENTS

ACKNOWLEDGEMENTS

WE EXTEND OUR SINCERE GRATITUDE TO THREE ORGANIS-ations, without whose help this book could not have been produced—Fortescue Metals Group Limited, The City of Melbourne and Outback Trees of Australia. Special thanks to Andrew and Nicola Forrest, Graeme Rowley, Professor Rob Adams and Denis O'Meara.

A huge thankyou to those involved in the production of the book, especially to Jenny Happell and Anne Findlay for their highly professional proofreading and editing skills. Thanks to The Miegunyah Press, especially to Tracy O'Shaughnessy and Cinzia Cavallaro, as well as to Hamish Freeman from Pfisterer & Freeman.

Many people around Australia have been exceedingly helpful, either providing advice about the book or recommending trees to be covered. Many were happy to meet us on dusty and remote roads to take us to their favourite trees. Others fed us and provided accommodation. We encountered nothing but enthusiasm from everyone involved in the book. Trees bind people together.

We are grateful to the following:

Bronwyn Alcorn, Jenny Allen, Judy Baillieu, Richard Barley, Hugh Beggs, Pat Bell, Chris Betteridge, Alex Bicknell, Rod Bird, Graeme Black, Pete Boyles, Warren Boyles, Stan Bracchi, David Burton, Ben Byrne, Sean Cadman, Dean Cameron, John Carswell, Mary Cassini, Carrick Chambers, Bruce Chisholm, Keith Corbett, Jane Cotter, Peter Crettenden, Gary Crockett, David Cunningham, Abby Desreaux, Deb and Anthony Desreaux, Norm Dunn, Paul Edwards, Tim Entwisle, Jason-Jay Fletcher, Michael and Helen Gannon, Sally Garrett, Brett Galloway, Jennifer Gardner, Pat Garratt, Giles Gibson, Gordon Glenn, Trevor 'Wing' Hagger, George Haig, Ebony Hall, Kerry Hanrahan, Heidi Harrison, Ronnie 'The Texmanian' Harrison, Kerry Hay, David Herbig, Kylie Hodge, Dennis Ihle, Jeff Ihle, Greg Ingram, Graeme Jephcott, Bomber Johnson, Wyn Jones, Geoff Kingston, Peter Krause, Pauline Ladiges, Geoff Law, John and Sandy Learmont, Daan Loock, Emma Lynch, Alan and Margaret McCallum, Fiona McCarthy, Alan Macfadyen, Ross McKinnon, Hugh and Fiona MacLachlan, Dougal and Angela McQuie, Butch Maher, Arika Maloney, Neville Marchant, Werner and Danielle Marschalek, Anne Marshall, Kiah Martin, Andrew Mendelawitz, Rudi Michelson, Brett Mifsud, Wayne Millen, Greg Moore, Bes Murray, Val Murray, Judymae Napier, Pam Nguyen, Lesley Nicklason, Paul Nolan, Kathleen Noonan, Roderic and Kate O'Connor, Mic O'Neil, Andrew Onley, Natalie Papworth, Heather Peel, John Perrignon, James Plant, Jaime Plaza, Moo Price, Delphine Puxty, Michael Ramsden, Sarah Rees, Jeannie and Peter Reynolds, Vanessa Richardson, Ron Riley, Sally Romanes, John Ross, Andrew Rouse, Ben Romcke, Ted Rudge, Bruce Saxton, Annabel Shears, Ian Shears, Ian Smith, Steve Strike, Martin Summons, Neil Teague, Kevin Thiele, Jeremy Thomas, Roger Underwood, Stephen van Leeuwen, Nick Viney, Matt Wallace, Gibb Wettenhall, David Wilkinson, Glenn Williams, Cliff Winfield, Adrian Woollard, Simon Wright, Simon and Brooke Yates.

The photograph on page 176 of Mr Bob Gunther, manager of *Monkira Station* in Queensland, and the giant coolibah, taken by Arthur Groom (1904–1953) in 1952 (nla.pic-an23182114), is reproduced by kind permission of the National Library of Australia. The photographs on pages 92–7 are reproduced by kind permission of Botanic Gardens Trust/Jaime Plaza.

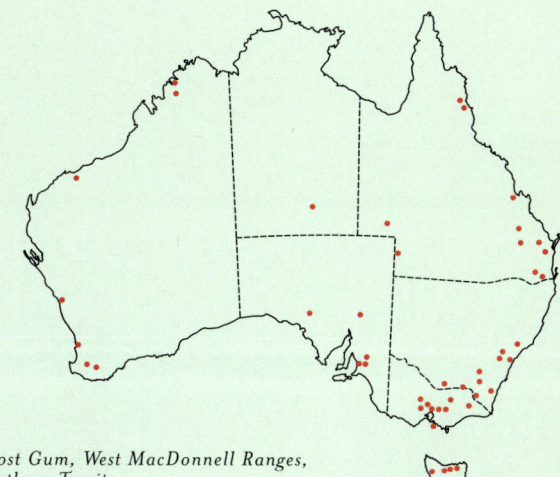

— left — Ghost Gum, West MacDonnell Ranges, Northern Territory

ON WEDNESDAY 24 OCTOBER 2007 PHOTOGRAPHER KIM Baker and I met with Kiah Martin, Australian tree-climbing champion and senior arborist at Melbourne's Royal Botanic Gardens, to chat about a common passion—trees.

Kiah walked us from Birdwood Avenue to her favourite tree—a stately and spectacular Algerian Oak (*Quercus canariensis*) known in the gardens as the Lady Loch Oak. The tree, planted in 1889 by the wife of Victoria's then governor Sir Henry Loch, occupied a prime position on the gardens' Oak Lawn with views to the north, over the city skyline. The 118-year-old tree, Kiah told us, was hugely popular with garden staff and members of the public alike.

We sat on the oak's sweeping branches, some of which were so long—up to 20 metres—that they touched the ground. We stroked the tree's rough bark, examined its leaves and gazed up through its 43-metre canopy, which filtered the dazzling spring-day light.

Kiah gave us the tree's history and talked about its longevity. She expressed admiration at how well this European tree had grown in the foreign Australian soil. She told us that some Algerian Oaks in Europe have lived for hundreds of years and consequently this tree, less than 120 years old, could well be still in its early stages of life.

'But,' she warned, 'the reality is that we don't really know how long these oak trees live in Australia because none are older than European settlement. Australian and European climates are very different. Today's drought is probably not helping. We will see.'

We resolved that the iconic Lady Loch Oak—with its interesting history, imposing size and shape, and its place in the hearts of so many people—deserved a spot in this book.

Without warning, less than a month later, at 10.30 a.m. on 15 November—the 118th anniversary of the day the governor's wife took a spade and planted the oak sapling—the Lady Loch Oak split clean down the middle and crashed to the ground. People were devastated. Some stood at a distance and stared in disbelief, others put bunches of flowers at its mangled base. It took several weeks to clear away the debris, the timber sold to boat builders and furniture makers.

The experts said the tree was a victim of the city's prolonged drought that had browned the springtime lawns of Australia's Garden State capital. The drought had also caused the city's stately English Elms (*Ulmus procera*) to shed their leaves early, in a desperate bid for survival.

The sudden and unexpected demise of the Lady Loch Oak reinforced, for us, the importance of this book.

The truth is that we tend to take our trees for granted, despite the fact that their lives are finite. Ancient trees can take centuries to grow, but can disappear—as the Lady Loch Oak amply demonstrated—in the blink of an eye. Hectares of them can vanish in an afternoon when chainsaws and bulldozers are involved. Thomas Hardy's

THE TIMBER INDUSTRY AND THE
ENVIRONMENTAL MOVEMENT
DEBATE THE FIGURES, BUT,
BROADLY, THEY AGREE THAT
ONLY 8 PER CENT OF AUSTRALIA'S
FOREST TODAY IS OLD-GROWTH
FOREST AND THAT FIGURE IS
FALLING EVERY YEAR.

poem, written a century ago, is a poignant stirrer of our ecological conscience.

The purpose of this book is to recognise our iconic trees, and the great forests in which they live. Trees are, after all, our largest and oldest living things. They are Australia's natural, national treasures—the true Elders of our vast continent.

On the face of it, things have come a long way in Australia since 1884, when brothers Bill Cornthwaite, a Victorian farmer, and George Cornthwaite, a government surveyor, were convinced that a regal Mountain Ash (*Eucalyptus regnans*) at Thorpdale in Gippsland was the world's tallest tree.

They reasoned there was only one way to prove it—cut it down and measure it. Which they did. And they were right. The tree was 114.3 metres tall (or long, by this stage) and the brothers claimed the record. Today a rusting plaque marks the spot of possibly Australia's most senseless bit of ecological vandalism.

Then again, perhaps things have not changed as much as we might think, or hope. Although Australia has the sixth-largest area of forest cover in the world—164 million hectares, or 21 per cent of the total land area, we have a fairly dismal record for forest clearing, and among the highest rate of species extinction and decline in the world.

The timber industry and the environmental movement debate the figures, but, broadly, they agree that only 8 per cent of Australia's forest today is old-growth forest and that figure is falling every year.

From 1983 to 1993 more than 5 million hectares of forest and woodland in Australia were cleared. Since then the rate of clearing has increased (although Queensland has recently officially stopped land clearing, which has helped the situation dramatically). Today, only 16 per cent of our total native forest area is protected. Victoria has lost half its old-growth forest. New South Wales has lost more. Sadly, 15 million hectares of forest in Australia (mainly New South Wales, Victoria and Tasmania) are today subject to commercial logging. Eight generations—less than 250 years—of European settlers have wrought much destruction on Australia's natural resources.

Industrial logging started in the 1940s in Australia's most tree-covered state, Tasmania, and it was only in the late 1990s that policies were developed to protect tall trees. Today, less than 20 per cent of Tasmania's tall old-growth forests remain—in spectacular places such as the Styx, the Florentine, the Weld and the Tarkine—and only about half of what is left is properly protected.

This is a book about old, large and fascinating trees, both native and introduced. Compiling it required travelling to all corners of a land that is singularly blessed with such trees. Some species, like the Huon Pine (*Lagarostrobos franklinii*) in Tasmania, live for thousands of years. The mighty Mountain Ash (*Eucalyptus regnans*), found in many parts of Tasmania and Victoria, is the world's tallest hardwood, and the second-tallest tree in the world after the Coast Redwood (*Sequoia sempervirens*), a conifer in the United States.

The Mountain Ash has several tall soulmates. Four other eucalypts (*E. regnans*, *E. globulus*, *E. viminalis* and *E. delegatensis*) are in the top-ten tallest tree species in the world. Two of the next five tallest are also Australian native species—*E. obliqua* and *E. nitens*. Many of the tallest of this impressive list of Australian trees are more than 500 years old, and have survived fire, drought, erosion and all manner of pestilence.

Remarkable trees can, of course, be both old and tall, or simply massive, like an extraordinary Green Fig on the Atherton Tablelands, which has a girth of more than 40 metres, or the gargantuan Boab (*Adansonia gregorii*) found on *Mount Hart* station in the Kimberley region of Western Australia, which measures 17 metres around its base. Countless generations of Aboriginal people have camped next to it, sheltered in its shadows and lived off its fruit.

Trees can be great survivors, like the ancient Red Tingle (*Eucalyptus jacksonii*) in the Walpole–Nornalup National Park in Western Australia, the trunk of which has been hollowed out by fire to the extent that it is hard to believe that the tree is still standing, yet it clings to life with admirable tenacity.

The resilience and resourcefulness of trees—allowing them to adapt and survive—continues to delight and amaze us. When the long-lost Wollemi Pine (*Wollemia nobilis*) was discovered in a remote canyon near Sydney in 1994, director of Sydney's Royal Botanic Gardens Professor Carrick Chambers announced that it was the 'equivalent of finding a small dinosaur alive on Earth'. It was reported around the world. The species had been regenerating in this small area for millions of years, surviving raging bushfires and no less than seventeen ice ages.

When it comes to adaptability, look no further than the River Red Gum (*Eucalyptus camaldulensis*) that grows south of Geraldton in Western Australia. The wind blows off the nearby Indian Ocean, covering the tree with salt. The tree has not given up, but it has reduced its growth rate and keeled over low to the ground in order to cling to life.

Compare this tree to the huge River Red Gum near Wilpena Pound in South Australia's Flinders Ranges, made famous by photographer Harold Cazneaux in 1937. The two trees are the same species but have completely different shapes and sizes, adapting differently to vastly different environments.

Trees have witnessed important parts of Australia's history, like the Coolibah (*Eucalyptus coolabah*) on Cooper Creek in Queensland, under which a cache of provisions was buried for ill-fated explorers Burke and Wills. Or the Boab on Prince Regent Sound in the Kimberley region of Western Australia on which, in 1820, seaman Phillip Parker King carved his boat's name, *Mermaid*, as evidence that he had landed there. The evidence remains to this day—possibly Australia's oldest remaining graffiti from the days of European exploration.

Our great trees have witnessed much, and endured much, and they have remarkable stories to tell. 'The best friend on Earth of man is the tree,' said American architect, writer and educator Frank Lloyd Wright. This is true not just because we can appreciate the beauty, elegance, majesty and serenity of trees, but because we can learn so much from them.

Trees are our friends because they provide shelter, for humans, animals and insects. Trees produce wood for houses, boats, tools and paper, they provide a carbon sink and create oxygen, which we breathe. As global warming becomes more of an issue on Earth, trees will become commensurately more important. It is increasingly likely that current woodchip royalties will become less valuable to us than living trees as carbon sinks to balance our future world trade.

According to ancient mythology, trees link the Earth to the sky. In this respect trees link humans to another world. Trees exude calmness, and humans can access that calmness. Where better to collect one's thoughts than beneath a tree? Or in a forest?

Great trees, and the ancient forests in which they live, should be recognised and protected for many reasons, principal of which is that they are important to humanity. Nearly 200 years ago, American nature writer Henry David Thoreau wrote, 'in wildness is the preservation of the world'.

Senator Bob Brown, in the foreword to the book *Endangered—Tasmania's Wild Places*, says, 'Wilderness has become one of the world's fastest disappearing resources, and it is non-renewable. Yet unlike oil, gold or woodchips, it is essential to the wellbeing of humanity. We are made of it and fashioned by it … our psychological beings resonate with it.'

'In the future of humanity, and of all the world in all its aspects, trees are key players,' says Colin Tudge in his book *The Tree*. 'It is also true—marvellously and encouragingly so—that societies can build their entire economies around trees; economies that are much better for people at large, and infinitely more sustainable, than anything we have at present. Trees could indeed stand at the heart of all the world's economics and politics, just as they are at the centre of all terrestrial ecology … In the future of humanity, and of all the world in all its aspects, trees are key players.'

Richard Allen

MAGNIFICENT
NATIVES

SNAPPY GUM

Eucalyptus leucophloia

Millstream–Chichester National Park
Pilbara, Western Australia

FOR CENTURIES THE YINDJIBARNDI people of what is now the Pilbara region of Western Australia gathered at Ngarrari, a sleepy waterhole 150 kilometres from the coast. They daubed their bodies in white paint and danced with abandon, hollering and stamping their feet in the red dust. They told fanciful stories of mythical beasts and feasted on roots and fish caught in the pools of the Fortescue River.

In the 1860s the region around Ngarrari was declared a pastoral lease. In its heyday *Millstream Station*, run by the formidable Gordon family, ran more than 55 000 sheep. In 1986 an allocation of 200 000 hectares was set aside for the Millstream–Chichester National Park.

In recent years the dusty tracks of this remote region have been sealed, and gleaming locomotives now pull carriages laden with iron ore from nearby mines to the coast, bound for China. Today visitors travel to the national park by air-conditioned four-wheel drives. Those few members of the Yindjibarndi people who do visit their old waterhole do so today as guides and employees of the park.

Witnessing all this activity from high on an escarpment close to the old *Millstream* homestead, possibly wondering about the meaning of it all, is an old Snappy Gum.

This tree has endured at least 200 searing summers but it does not look any the worse for it; it looks like it could do another two centuries without a problem. Its two shiny and twisted trunks, knotted and warty, support a healthy canopy of leaves.

Snappy Gums are the Pilbara's most iconic tree. They are not large—this one is no more than 6 metres tall—but they are content in hot, dry country. They are happiest clinging to sheer cliff faces, and their white, almost translucent, trunks (*leucophloia* means white bark) can be seen for miles as they eke out an existence on the featureless rocky plains.

And there are few places in Australia more forbidding than the Pilbara region. Temperatures regularly top 45 degrees and the four months from August to November rarely yield more than 10 millimetres of rain. The tree does not have much company; only a handful of smaller Snappy Gums and some Bloodwoods (*Corymbia hamersleyana*), surrounded by a sea of spinifex.

To the south, 100 kilometres away, is the Hamersley Range, and each night it turns a symphony of blues and purples in the setting sun. Closer to the Snappy Gum, at the bottom of the escarpment on which it sits, is Crossing Pool, which is fed by an underground aquifer and overlooked by stately River Red Gums (*Eucalyptus camaldulensis*) and Silver-leaved Paperbarks (*Melaleuca argentea*).

SNAPPY GUMS SURVIVE IN TEMPERATURES THAT REGULARLY TOP 45 DEGREES AND THE FOUR MONTHS FROM AUGUST TO NOVEMBER RARELY YIELD MORE THAN 10 MILLIMETRES OF RAIN.

TUART
Eucalyptus gomphocephala

Kings Park and Botanic Garden
Perth, Western Australia

TWO HUNDRED YEARS AGO VAST forests of Tuart, Jarrah and Marri stretched along the West Australian coast from what is now Jurien Bay to near Busselton, a distance of 350 kilometres. Today only a quarter of the original Tuart forests and woodlands remain. This is due mainly to land clearing by the early European settlers who farmed and logged this fertile tract. Some Tuart forests fell prey to disease.

Tuart timber, a very dense hardwood, was used to build the city of Perth and its surrounding townships, and to support vast mining and farming enterprises. The wood was fashioned into wagon wheels, decking for wagons, bridge supports and tool handles. In more recent times the timber has been used for furniture and house flooring.

In the twenty-first century Tuart forest distribution is limited to a narrow coastal belt. In some areas these trees are found short distances inland, usually along river courses. Today the Tuart, which is not closely related to any other eucalypts, is largely protected.

Thankfully, the 400-hectare Kings Park and Botanic Garden, to the west of central Perth, is doing its best to preserve what remains of the city's largest native tree. It is home to hundreds of Tuarts. The biggest of them, 40 metres high and probably 100 years old, occupies a beautiful stretch of sun-drenched lawn, south of Forrest Drive, overlooking the Swan River.

Tuarts can live for up to 500 years, so this one is just beginning its long life—if, that is, it doesn't succumb to the scourge of *Armillaria luteobubalina*, a soil-borne fungus that causes roots to rot, and has been killing trees at Kings Park for decades.

Tuarts are the largest trees on the Swan Coastal Plain near Perth and provide an important habitat for animals and birds. They are easily recognised by their flower buds, which have swollen caps and look endearingly like small ice-cream cones (*gomphocephala* means 'club-headed').

TUARTS CAN LIVE FOR UP TO 500 YEARS, SO THIS ONE IS JUST BEGINNING ITS LONG LIFE.

SNOW GUM

Eucalyptus pauciflora ssp. niphophila

Bogong High Plains
Victoria

IN 1823, 34-YEAR-OLD CZECH BOTANIST FRANZ WILHELM Sieber came to Australia and industriously collected and dried cuttings from 645 Australian plants. He returned to Europe after only seven months and began the arduous task of naming and classifying his finds. He named one of the trees *Eucalyptus pauciflora*—*pauci* meaning 'few' and *florus* meaning 'flowered'.

Anyone who has been to the Australian high country and seen this tree, better known as a Snow Gum, in full flower—generally between December and February—will know that Sieber got it wrong. Snow Gums have beautiful white flowers, and lots of them. Perhaps it was a poor cutting that he took home, or maybe it was damaged en route. We will never know; the unfortunate man died in 1844 at the Prague lunatic asylum, but his legacy remains.

The Snow Gum—also known as White Sallee, Cabbage Gum, Weeping Gum or Ghost Gum (not to be confused with the Ghost Gum in central Australia)—is found primarily in the high country of New South Wales, Victoria and Tasmania. It is mostly found in altitudes above 1000 metres, and it is very long-lived and slow growing.

Unlike most other trees, which thrive best in benign sunny and moist conditions, the Snow Gum prefers a much tougher environment. That is why you see them in Australia's highest, coldest and windiest places, more often than not covered in ice and snow, battling blizzards and sleet.

In many of these places, Snow Gums are the only trees growing, their tortured limbs often spreading low to the ground. When they die, their white skeletons often remain for many years, a reminder of the difficult lives they have endured.

There is no better example of the Snow Gum than this monster near Kelly Hut on the Bogong High Plains in Victoria. From the ski resort of Falls Creek, take the road past Rocky Valley Dam and follow the Big River Fire Track to the south-east—it is worth the trek.

Like many other high-altitude Snow Gums, this tree's trunk is fantastically twisted and tortured—the product of many decades of bitter winters. Its bark is a maze of colourful ribbons of brown, gold, white and red. It is as though an artist is trying out stripes of different colours and just can't come to a decision.

We photographed the tree on a sunny spring afternoon. It shifted about in the wind, looking restless, as if waiting for the next cold snap. As we ended our shoot, a cold front was moving in and we retreated. The tree looked content.

MOONAH
Melaleuca lanceolata

Churchill Island
Westernport Bay, Victoria

IN 1798, WHEN ENGLISH EXPLORER George Bass—naval surgeon and enthusiastic naturalist and botanist—sailed in a whaleboat from Sydney to what would later become Victoria, he decided to follow a route around the northern side of what is now Phillip Island in Westernport Bay. He and his crew of six could not have helped but notice a stand of tall, twisted trees against the skyline on a small island to the north of Phillip Island.

These same trees, Moonahs, still exist on the small 57-hectare Churchill Island. It was named by Lieutenant James Grant in 1801, who sailed there in his ship *Lady Nelson*. Grant thought the island a good place for a garden and named it after Peter Churchill of Devon, who gave Grant seeds to plant in Australia for 'the future benefit of our fellow men'.

For thousands of years, before either Bass or Grant saw the island, the local Bunurong people used to paddle to what they called Moonah Mia in dugouts or canoes made from bark. There they caught fish, stingrays, seals and muttonbirds, and collected shellfish, bulbs and red ochre, a pigment they used in ceremonies. They called the strange, twisted trees on the island Moonahs.

Estimates are that this Moonah on Churchill Island, 150 kilometres south-east of Melbourne, is more than 350 years old. It sits high on the brow of a hill rising from the ocean, and has witnessed many changes in Westernport Bay in the two centuries since the energetic Bass visited. The history of European settlement has floated past this venerable tree—including the arrival of sealing boats, bark cutters, explorers, scientists, surveyors and, more recently, settlers and holidaymakers.

Moonahs are one of around 140 species of melaleuca that stretch across the southern half of Australia. Those who believe it is predominantly an east coast tree might be surprised at its other common name—Rottnest Island Tea-tree. Moonahs are happy living in sandy soil, often in windswept areas close to the ocean. They are distinguished from many of the other species by their dark, almost black, corrugated bark (unlike the 'paperbark' melaleucas) and small dark green lance-shaped leaves. Creamy bottlebrush-like flowers grow in clusters on the older branches and the fruits are small globular capsules.

FOR THOUSANDS OF YEARS, BEFORE EITHER BASS OR GRANT SAW THE ISLAND, THE LOCAL BUNURONG PEOPLE USED TO PADDLE TO WHAT THEY CALLED MOONAH MIA IN DUGOUTS OR CANOES MADE FROM BARK.

QUEENSLAND BOTTLE TREE

Brachychiton rupestris

State Forest
Burnett region, Queensland

PEOPLE AROUND GAYNDAH, IN THE BURNETT REGION OF Queensland, 150 kilometres west of Maryborough, have a name for this Queensland Bottle Tree. They call it, simply, 'the big one'.

The tree grows in a remote State Forest not far from Crooked Creek Road, about four hours' drive north of Brisbane. At 23 metres tall, with a 9.5-metre circumference, it is both immensely broad and imposing, towering over its close neighbours—other smaller Queensland Bottle Trees, Kurrajongs, and an impressive spreading White Cedar.

In many ways the tree is lucky to be alive at all. The region used to be dry rainforest, but was cleared—by axe—in the 1930s, then again in the 1960s by bulldozer and chain. Land clearers were not known for either sentiment or clemency.

Death by dozer is not the only danger the tree has overcome. Half a dozen other Queensland Bottle Trees in its paddock have been hit by lightning in recent years, splitting them clean down the middle. One such tree was only 45 metres from 'the big one'.

Early settlers used to spare these pudgy trees, keeping them not only for their appealing shape but as a source of fodder in times of drought. Cattle make short work of fallen Queensland Bottle Trees, eating both the leaves and the pithy centre of the trunk. The rest rots away in the heat.

Parks and wildlife officers have done their best to make sure that this tree doesn't fall victim to that other bush scourge—fire. Nineteen years ago a firebreak was cut around the tree, and it remains to this day.

Queensland Bottle Trees look very similar to the Boabs of the Kimberley region of Western Australia. Both have large, bloated trunks and short, spindly branches. But they are not related. Queensland Bottle Trees belong to a different genus, *Brachychiton*, and are related to Illawarra Flame Trees, Lacebark Trees and Kurrajongs. The Queensland Bottle Tree is endemic to central Queensland and northern New South Wales. There are some thirty *Brachychiton* relatives in Australia and they tend to grow further south.

The seeds, roots, stems and bark of the Queensland Bottle Tree have traditionally been sources of food for people as well as animals. Its lacy inner bark was an important source of fibre for Aboriginal people who made twine and dilly bags from it. Some enterprising individuals even wove it into fishing nets.

We can only guess how old this tree is. Queensland Bottle Trees don't take kindly to core testing to determine their age. Such coring can kill the tree. A bulldozer accidentally nicking or bruising the tree's bark can do the same. A farmer who lives near 'the big one', and who has kept an eye on it for decades, says it has grown in diameter by only 10 centimetres in the past fifty years.

The average annual rainfall around these parts is 650 millimetres. The past two years have yielded only 870 millimetres. Consequently, the trees in the region are under some pressure and some older eucalypts have turned up their toes. 'The big one' is hanging in there; it has adapted perfectly to its harsh environs.

BUNYA PINE

Araucaria bidwillii

Bunya Mountains National Park
Queensland

But whether urged by love or hate
By the fair prospect made elate,
The tribes in battle celebrate
The last great Bunya Feast.

The Feast of the Bunya, Cornelius Moynihan

THIS STATUESQUE AND BEAUTIFUL BUNYA PINE SITS JUST OFF Stirling Way, on the eastern edge of the Bunya Mountains National Park in south-east Queensland. It has one of the country's best views, 100 kilometres to the east you can see the Jimna Range.

Although it has some company—other Bunyas appear above the rainforest canopy nearby, towering over Red Cedars, Figs, Piccabeens and Giant Stinging Trees—to all intents and purposes it sits in glorious isolation. Its main visitors are King Parrots, Paradise Riflebirds and Crimson Rosellas, which swoop madly in and around its branches.

Although in ancient times the Bunya Pine covered much of eastern Australia, today it is relatively scarce. Being an excellent timber tree, it suffered intensive logging in the 1860s. It is endemic to only three places: the Bunya Mountains, the Blackall Range west of Caloundra, and Mount Molloy west of Port Douglas.

The Bunya Pine was first recorded by a non-Indigenous Australian in the 1830s. He was Moreton Bay's Superin- tendent of Works, Andrew Petrie, and the tree became known as Petrie's Pine. Its botanical name was adopted after botanist John Bidwill sent specimens to London's Kew Gardens. The arborists there were suitably impressed and called it *Araucaria bidwillii*.

In fact this tree is not a pine, but a member of the family Araucariaceae. Worldwide there are thirty-eight species of this family (in three genera, *Agathis*, *Araucaria* and *Wollemia*), and all three genera are represented in Australia. Aside from the Bunya Pine, there are only two other native Australian *Araucaria*, the Hoop Pine (*Araucaria cunninghamii*) and the Norfolk Island Pine (*Araucaria heterophylla*). Of the nineteen species of *Araucaria* worldwide, thirteen are endemic to New Caledonia.

All are ancient trees. Fossil records show that the Bunya Pine and the Hoop Pine replaced ferns as the major vegetation about 200 million years ago. The Bunya Pine provided shelter, and probably food, for dinosaurs. The tree certainly looks prehistoric, with long, pendulous branches and a straight trunk covered in rough, pock- marked bark. Its leaves are sharp, and resemble the spines on the back of a stegosaurus.

It is surely one of Australia's most dangerous trees. Not only can its leaves make a sharp impression on a hand or foot but its cones—which can grow to the size of a football and can weigh more than 10 kilograms—can be lethal when dropping from the high branches in late summer. Walk through a Bunya forest at your peril!

The Bunya Pine is monoecious, meaning each tree has both male and female cones. It releases pollen around September, and fifteen months later its cones are ripe.

THE BUNYA PINE PROVIDED SHELTER, AND PROBABLY FOOD, FOR DINOSAURS. THE TREE CERTAINLY LOOKS PREHISTORIC, WITH LONG, PENDULOUS BRANCHES AND A STRAIGHT TRUNK COVERED IN ROUGH, POCKMARKED BARK. ITS LEAVES ARE SHARP, AND RESEMBLE THE SPINES ON THE BACK OF A STEGOSAURUS.

Bounteous seasons generally occur every three years. The last great season was in 2007.

For thousands of years the Bunya Pine was one of few Australian trees to dramatically influence human behaviour. In 1846, members of Ludwig Leichhardt's exploration team noticed a long line of Aboriginal people picking their way through scrub in eastern Queensland, heading south. They were thin—indeed some were emaciated—but, as they went, they laughed together and walked with purpose.

Daniel Bunce, a botanist on the expedition, wrote in his diary that the people were 'on their way to the Bunya Bunya country for the purpose of obtaining a very valuable fruit, the product of the *Araucaria bidwillii*'.

After receiving message sticks, up to 2000 Aboriginal people at a time, from tribes including the Bigambul, Barungam and Mandanbanji, would regularly walk hundreds of kilometres from as far north as the Burnett River (today's Bundaberg), as far south as the Clarence River (today's Grafton) and as far west as the Maranoa River (today's Roma). Their destination was the 1100-metre Bunya Mountains, 160 kilometres northwest of present-day Brisbane, to feast on the nuts of what they knew as the *bonye bonye*.

Once they arrived at the foot of the mountains they set up camp and waited. When the Bunya cones ripened in late summer the Waka Waka people, who owned the trees (a very unusual concept among Aboriginal people),

would invite their guests to the top of the mountain. The hosts would then climb the trees with the aid of vines and toeholds in the bark and knock down the ripe Bunya cones. (Trees can produce fifty cones in a bumper year, some even more.)

The cones would be gathered and broken up. The nuts inside would be cooked and the milky flesh eaten with much rejoicing. The days were spent hunting, in pretend duels and sham fights. The nights were full of songs and dance, stories and riddles. The festivals—and that's precisely what they were—would last for weeks.

Members of Leichhardt's expedition were unaware of it, but they were witnessing the precursor to one of the last great Bunya festivals. The final such festival was held in 1875. After that, white settlement made difficult the journeys of the Aboriginal people and the logging industry stripped away many of the trees.

Following the last Bunya festival Archibald Meston, botanist and journalist, wrote:

Alas, no artist ever beheld those strange scenes at the assemblings of the tribes at the Bunya Mountains. They have gone for ever; vague and shadowy now in the misty moonlight of memory, dim phantoms only in the imagination. In fancy alone can we recall those multitudinous dark forms, stalking stealthily through the pine scrubs; in fancy only can we hear the soft footfalls of a thousand naked feet upon the fallen leaves.

QUEENSLAND GREY IRONBARK

Eucalyptus drepanophylla

Jimna State Forest
Queensland

TWENTY-SIX KILOMETRES NORTH of Jimna in Queensland, and a bone-shaking drive off the dusty and corrugated Kilcoy–Murgon Road, is a mighty and quintessentially Australian tree.

This old Ironbark takes a bit of getting to—enter through a gate with *Wombalinger* painted on a bit of old paling, and then drive 3 kilometres along a bouncing dirt track, turning left at the fork. But the trip is worth it. The tree grows on a crest in a small clearing, not far from Kingaham Creek. It is tall and beautifully proportioned, its trunk as straight as the proverbial gun barrel.

Ironbarks are one of Australia's most famous and evocative trees. Unlike many other eucalypts, they do not shed their bark each year. The dark grey bark accumulates, providing hardy, deeply furrowed fire-resistant protection. Hence their part in Australian folklore, celebrating strength, robustness and a never-say-die attitude. Little wonder that Banjo Paterson chose the Man from Ironbark as his model when looking for a simple country bloke who took no cheek from a mischievous Sydney barber.

This tree, nearly 6 metres in circumference, can probably count its blessings. The area in which it lives, lot 986 in the Jimna State Forest, was logged in 1987, providing wood for sawmills in nearby Kilcoy, Yednia and Woodford. The mills feasted on local Tallowwood, Gum-topped Box, Yellow Stringybark, Brush Box and, of course, Ironbark. Ironbark was valued for its strength and durability and used for wharves, bridges, telegraph poles, girders and fence posts.

Perhaps a tree feller became a touch emotional when he looked at this tree and decided to spare it. However it came about, two years ago the area was declared a National Park and the old Ironbark was saved. Today four white pegs show that it is a protected site of beauty. The only things likely to bother it now are the drought and cattle rubbing up against its bark. Not that they would make much of an impression.

IRONBARKS ARE PART OF AUSTRALIAN FOLKLORE, CELEBRATING STRENGTH, ROBUSTNESS AND A NEVER-SAY-DIE ATTITUDE.

TASMANIAN MYRTLE

Nothofagus cunninghamii

Paradise Plain
North-east Tasmania

MYRTLE BEECH ARE ROMANTIC evergreen trees found in moist mountain valleys. They often have luscious, thick coverings of lichen and moss, giving them a regal and vestmented appearance.

They were once very common in Victoria's Otway Ranges, Baw Baw mountains and in the Yarra Valley but, as with many trees whose timber is easy to work with and polishes well, the vast majority of Myrtle Beech in these areas have disappeared. Fire, too, has taken its toll, and today they are hard to find in Victoria.

Thankfully, they are still common in Tasmania, where the locals refer to them as Tasmanian Myrtle.

This 30-metre-high specimen—situated at an altitude of nearly 1000 metres on the edge of Paradise Plain, south of the farming district of Ringarooma 100 kilometres north-east of Launceston—is a startlingly beautiful tree. Seven or eight mighty branches loop downwards, some touching the ground before shooting skywards again.

It has every reason to be healthy, and happy. Its surroundings are ideal, including a bubbling creek nearby.

The forest around it is damp, which myrtles love, and the tree lives on the edge of a wide plain of native grasses. Other native Tasmanian trees have taken advantage of the ideal growing conditions, including Southern Sassafras, Celery-top Pines and Mountain Peppers.

Like many Tasmanian Myrtles, this tree plays host to other plants. Blue circles of lichen run up the underside of the tree's scaly branches, and large clumps of moss sit happily near its crown. Kangaroo Paw ferns (*Microsorum diversifolium*) glide up the branches like rampant boas.

Tasmanian Myrtles are not deciduous, so this tree has a permanent canopy of leaves. New growth arrives in the springtime, the leaves turning from lime green to yellow, then orange and red. Small flowers appear in early summer; the male flowers appear on the lower branches and the female flowers occur in threes higher up.

The tree can, in a way, count itself lucky. The banks of Paradise Hill, only 500 metres away, have been logged extensively. The steep slopes around this tree, fortunately for it, are not suitable for plantation, so this fern-clad matriarch is likely to enjoy good health for many years to come.

LIKE MANY TASMANIAN MYRTLES, THIS TREE PLAYS HOST TO OTHER PLANTS. BLUE CIRCLES OF LICHEN RUN UP THE UNDERSIDE OF THE TREE'S SCALY BRANCHES, AND LARGE CLUMPS OF MOSS SIT HAPPILY NEAR ITS CROWN.

MORETON BAY FIG

Ficus macrophylla f. macrophylla

Royal Botanic Gardens
Sydney, New South Wales

CHARLES MOORE, LIKE MANY other heads of Australia's botanic gardens, was undoubtedly committed to his task. For nearly half a century, from 1848 to 1896, he ran Sydney's Royal Botanic Gardens.

History has it that it took Sydneysiders some time to accept him. He was seen as responsible for ousting the respected John Bidwill (after whom the Bunya Pine, *Araucaria bidwillii*, was named), and some were inclined to dismiss him because he had little formal training as a botanist. Many viewed him more as a gardener.

Moore, a native of Dundee, Scotland, was not deterred. He carried on the work of his predecessors and planted many subtropical trees in the gardens, such as Kauri Pines. But his favourite was the Moreton Bay Fig, a strangler fig found in rainforests from southern Queensland to the Illawarra scrubs in New South Wales. Moore planted hundreds of them, relishing their toughness, adaptability, and the shade provided by their vast canopies. Some say he went over the top with his plantings. Certainly Sydney today has no shortage of Moreton Bay Figs.

Details of what Moore planted in the gardens, including the dates of plantings, are a little sketchy. His successor, Joseph Maiden, regretted that Moore 'did not commit to paper the horticultural and botanical reminiscences of his long official career … His dislike of writing extended even to letter writing.' His place among Australian botany is assured, however; Melbourne Botanic Gardens director Ferdinand von Mueller named nineteen species after him.

Probably Moore's most famous Moreton Bay Fig grows 50 metres from the clear waters of Farm Cove in the northern section of the gardens. For decades it has been a favourite with children, who have picnicked in its shade, scaled its vast branches and explored the caves hidden among its cable-like roots.

The Fig was named the Children's Tree in 1983 and pupils from Plunket Street School were designated custodians of the tree.

Today the Children's Tree, 15 metres in circumference at chest height with a canopy of nearly 50 metres, is a picture of strength and resilience. It is hard to believe the trunk is strong enough to hold its enormous gravity-defying branches, which surely weigh several tonnes each. One giant bough, the best part of 30 metres long, touches the ground. No doubt the tree gains much strength from its vast roots, which fan out from the base like oversized serpents and serve as buttresses to the overworked trunk.

Like many old trees, this one could soon become dangerous. Figs can drop limbs unexpectedly, especially on still summer days. For this reason many of the figs in the Royal Botanic Gardens have been pruned extensively.

CHILDREN AND TREES ARE A GREAT MATCH AND ARE ALIKE IN MANY WAYS. GIVE THEM SPACE AND CARE AND THEY WILL REACH THEIR FULL POTENTIAL.

But due to the space around this specimen, it has been able to retain much of its natural form. Ten years ago the gardens' management decided to erect a fence around the tree, to protect visitors and employees. Today it shares its enclosure with another Fig (a multi-stemmed form of *Ficus macrophylla* from Lord Howe Island) and a giant Mexican Bald Cypress (*Taxodium mucronatum*).

'The Fig will probably lose a limb or two in the years ahead, but we will allow it to age gracefully,' says Tim Entwisle, the current executive director of the gardens. 'Some older trees in the gardens' high-use areas need to be heavily trimmed or replaced, but we made a conscious decision to allow this one to live out its life. It is in great condition, which means it will last longer. It really is a much-loved tree.'

Some trees in the Sydney Botanic Gardens survive from when the gardens were established in 1816; they are the oldest botanic gardens in Australia and the second oldest in the Southern Hemisphere after those in Rio de Janeiro. The 35-hectare gardens, which run alongside Sydney's famous harbour, are a quintessential part of the city—a favourite area for runners in the mornings and evenings, sunbathers and picnickers during the day and city workers at lunchtime.

The Children's Tree will continue to grace the gardens for some decades yet, allowing children to continue to wonder at it, and adults to reminisce. Tim Entwisle says children and trees are a great match: 'In fact they are alike in many ways. Give them space and care and they will reach their full potential.'

GHOST GUM

Corymbia aparrerinja

West MacDonnell Ranges
Northern Territory

THE CELEBRATED ABORIGINAL painter Albert Namatjira loved the Ghost Gums of the Northern Territory, and it is easy to see why. They are evocatively Australian, their white trunks contrasting with the red earth and deep blue sky of the Dreamtime region that has for centuries sustained Namatjira's Aranda people.

Australian writer Murray Bail wrote in his novel *Eucalyptus*, 'As for the Ghost Gum … there are those who maintain with a lump in their throats it is the most beautiful tree on earth.'

Ghost Gums, also known as Desert Gums, are solitary trees. It is not unusual to see one standing on its own, miles from its nearest neighbour. They are hardy too, seemingly just as happy growing from the side of sheer cliffs or rocky buttresses as on dusty plains and hidden gorges.

They do not travel widely. Most are found in central Australia—principally the Northern Territory—with a few in far eastern Western Australia and some in south-west Queensland.

Until recently the Ghost Gum was known as *Eucalyptus papuana*, but botanists recently changed the generic name of many eucalypts to *Corymbia*. The species name has changed too. *Corymbia papuana* is now reserved for the Ghost Gums of Papua New Guinea and Cape York. (The word *aparrerinja* is the central Australian Aboriginal word for Ghost Gum.)

If you take Larapinta Drive 40 kilometres out of Alice Springs, tracking the West MacDonnell Ranges, and keep an eye out as you approach the turn-off to Standley Chasm you won't miss this dramatic Ghost Gum—100 metres off the road on the left, standing proudly above the surrounding scrub. Its smooth lines sit beautifully against the stark and dramatic backdrop of the ranges.

This Ghost Gum is the embodiment of bloody-mindedness in the face of extreme odds. There can be few less hospitable places in the country, and yet this tree has endured the best part of 200 searing summers in the dry red earth. It is healthy and looks like it could do 200 more. A small dry creek bed cracks through the dusty soil 5 metres from its base, but nourishment would be rare. It does not rain much around here—less than 300 millimetres a year.

The tree's smooth white bark bears witness to a few struggles. A flame-like scar licks up its southern flank and charcoal marks indicate a battle with a fire in recent times. Its branches are scarred and bumped, like a witch's warty hand. A colony of ants has made its home from brown leaves in the tree's upper branches. They march, soldier-like, up and down the trunk in an endless procession of industry.

It could be said that Namatjira stumbled across his chosen craft. In 1936, aged thirty-four, he volunteered to show a painter from Melbourne, Rex Batterbee, good places to paint near Alice Springs in return for some lessons. Namatjira showed great aptitude and, two years later in Melbourne, his first exhibition of deeply detailed watercolours sold out. Exhibitions in Adelaide and Sydney were equally popular.

Albert Namatjira's most popular paintings are of Ghost Gums. Two such paintings appeared on Australian stamps, in 1993 and 2002.

THIS GHOST GUM IS THE EMBODIMENT OF BLOODY-MINDEDNESS IN THE FACE OF EXTREME ODDS. THERE CAN BE FEW LESS HOSPITABLE PLACES IN THE COUNTRY, AND YET THIS TREE HAS ENDURED THE BEST PART OF 200 SEARING SUMMERS IN THE DRY RED EARTH.

RED SILK COTTON TREE

Bombax ceiba var. leiocarpum

Rockhampton Botanic Gardens
Queensland

IN 1854, FOLLOWING THE POPULAR uprising in France which ended the reign of King Louis-Philippe and led to the creation of the Second Republic, a Frenchman—Anthelme Thozet—and his wife Maria Isabella fled their country as political refugees and sailed to Sydney.

Thozet, a keen botanist, took work at the Sydney Botanic Gardens but was lured to Canoona, 60 kilometres north-west of Rockhampton in Queensland, when gold was discovered there in 1858. He set up a grog tent in the goldfields and later built and ran the Alliance Hotel in Rockhampton.

But Thozet's first love was botany. He established an experimental garden, *Muellerville*, in North Rockhampton. He joined forces with a Scotsman, James Edgar, who had trained at Kew Gardens in London, and Melbourne Botanic Gardens director Ferdinand von Mueller to lobby the government to provide funds to build a botanic garden in Rockhampton.

The funds were eventually provided by the Queensland Acclimatization Society, whose mission was to import exotic species of fauna and flora for use in the new colonies. The money was enough to buy a horse paddock that was being used by the Aboriginal police. This was the start of the Botanic Gardens and planting began.

Thozet donated many tree seeds. Edgar became the gardens' first curator in 1873 and stayed thirty-two years. Only five curators have followed him.

Thozet and Edgar planted hundreds of trees. One of their passions was an experimental garden to assess which introduced species fared best in a tropical climate. Many trees still thrive in this garden today, including the bizarre South African Sausage Tree, Teak and Kapok trees from Asia, and a towering Bald Cypress from the Everglades in Florida.

Not far away, near the cenotaph in the middle of the gardens, is another of the gardens' first plantings—a spectacular Red Silk Cotton Tree. The 25-metre-high tree has a 40-metre canopy and its trunk a 6-metre circumference. The trunk is surprisingly narrow below the branches.

The tree's roots spread out above ground level, serving as buttresses, stabilising the tree. They have served it well because Rockhampton is regularly buffeted by severe storms from the west. Cyclones are not unheard of. If there are two such storms a year—a conservative estimate—then this tree has survived more than 200 of them.

The Red Silk Cotton Tree is native to deciduous vine forests of the Northern Territory and Cape York, and also Tropical Asia. It was quite common in

ITS SEED PODS ARE FULL OF FLUFFY FIBRE, LIKE RAW SILK, WHICH ASSISTS THE PODS TO FLOAT AND TO STAY MOIST, IMPROVING THE CHANCES OF THE SEEDS ACHIEVING GERMINATION.

THE ABORIGINAL PEOPLE USED TO
HOLLOW OUT THE BRANCHES TO
MAKE CANOES AND COOLAMONS
(BASIN-SHAPED DISHES).

Queensland until the 1940s, but became a victim of rapacious land clearing.

The tree flowers profusely in early spring and the nectar of its bright orange-red flowers is popular with Rockhampton's Rainbow Lorikeets. Its seed pods are full of fluffy fibre, like raw silk, which assists the pods to float and to stay moist, improving the chances of the seeds achieving germination.

The soft wood of the Red Silk Cotton Tree is not particularly useful, although the Aboriginal people used to hollow out the branches to make canoes and coolamons (basin-shaped dishes). The wood of other members of the Bombacaceae family is used, variously, for matches, floats and, not surprisingly, coffins. The tree is closely related to *Ochroma pyramidale*, which is native to tropical America, from Mexico to Brazil. Any child who has made model aeroplanes knows the tree, and its wood, by another name—balsa.

And what of Thozet? He died in 1878 aged fifty-two while collecting seeds west of Rockhampton. Eventually his estate went broke and the bank, in settlement, even repossessed his headstone. His grave in Rockhampton remains unmarked. But the Rockhampton Gardens, which were as much his vision as anyone's, make a fine substitute.

OLD CURIOSITIES

RED TINGLE
Eucalyptus jacksonii

Walpole–Nornalup National Park
Western Australia

ON 29 OCTOBER 1941 BRITISH PRIME Minister Winston Churchill famously implored the students at his alma mater, Harrow School: 'Never give in, never give in, never, never, never.'

This 400-year-old Red Tingle at Walpole–Nornalup National Park, 10 kilometres east of Walpole in the south-west of Western Australia, is the arboreal embodiment of Churchill's urgings.

It has lost the upper half of its trunk, and its higher branches—those that have not been obliterated by wind or lightning—are broken and battered. The inside of its vast trunk, the best part of 20 metres in circumference, has been eaten away by insects, fungus and fire.

But, courageously, it lives on, drawing nutrients and water from the ground via what is left of its vast trunk. It must be doing something right; small branches with fresh green leaves sprout from its highest points. Where there's life, there's hope.

The hollow trunk is like a Tolkienesque cavern, 15 metres high, one-third the height of the tree. One hundred people could quite comfortably throw a party there, with room to move.

Inside the cavern great slabs of wood, blackened by fire, hang down like giant internal buttresses. The huge entrance aside, small gaps in the trunk offer glimpses of the surrounding forest that is made up, for the most part, of other proud Red Tingles. Two other openings, extending 7 metres in height, look like the handiwork of a giant's axe.

The Red Tingle, the dominant tree in Walpole–Nornalup National Park, is very rare, and grows only in this region. It is the largest-based of all eucalypts, often having a massive butt with shallow roots spreading outwards to provide stability. Its timber, with similar properties to Jarrah, is deep pink to purplish in colour. Due to fires and now—fortunately—conservation, Tingle wood is seldom commercially available.

Importantly, Red Tingles have adapted to fire, their thick fibrous bark sustaining life even when the trunk has been hollowed out by bushfire and fungus. The last major fires in this region of Western Australia were in 1937 and 1951.

This tree is known as the Grandmother of Walpole–Nornalup National Park. It was put on the map in 1952 when Ted Bellanger, a descendant of one of the first European families to settle in the Walpole area, came across it when clearing a firebreak. Grainy photographs exist from that time of a large car backed into the cavernous trunk.

THE HOLLOW TRUNK IS LIKE A TOLKIENESQUE CAVERN, 15 METRES HIGH, ONE-THIRD THE HEIGHT OF THE TREE. ONE HUNDRED PEOPLE COULD QUITE COMFORTABLY THROW A PARTY THERE, WITH ROOM TO MOVE.

BOAB
Adansonia gregorii

Near *Mount Hart* station
South-west Kimberley, Western Australia

IS THERE SUCH THING AS A SPHERICAL TREE?

At the base of Junction Hill at the confluence of the Sprigg and Isdell rivers in Western Australia's lush Kimberley region, 165 kilometres north-east of Derby, is a Boab as close to spherical as you could find.

Its size is gargantuan, its shape phantasmagorical.

Boabs store moisture in their soft and fibrous tissue to help them survive the harsh dry season, which extends in the Kimberley from April to November. It is hard to believe any Boab could store more than this monster, which has a circumference of 17 metres and has clearly gorged on the good life for decades.

Boabs can live for more than 500 years, so this tree is probably centuries old. We will never know for sure; the age of old Boabs is difficult to determine as they are invariably hollow and have no annual growth rings.

The people in the Kimberley are a resourceful lot, and have put hollowed-out Boab trunks to various uses, including storing water. Some Boabs have been used for shops and for many years a hollow Boab south of Derby was used as a prison.

This Boab is hard to get to. It sits on unallocated crown land near the King Leopold Range and is either a five-hour hike from the homestead at *Mount Hart* station north of the remote Gibb River Road, or a one-hour helicopter ride from Derby.

That's not to say that the tree has been bereft of visitors during its long life. Thousands of Aboriginal people have doubtless sought its shade from the blistering summer sun or camped near its imposing bulk. A small flat stone—the size and shape of a paving stone—is embedded in the grass at the tree's base, probably used for Aboriginal tool-making or the crushing of seeds.

The nuts of the Boab drop off in May and June and are good bush tucker. The oval fruit, larger than an emu egg, contain many kidney-shaped seeds embedded in white pith that tastes like sherbet. The seeds have been likened to hazel nuts or almonds, and were ground up or eaten raw by the Aboriginal people. Both pith and seeds are rich in minerals. The roots of young Boabs can also be eaten. Humans are not the only ones to recognise the Boab as a source of food; cattle make short work of most parts of a fallen tree.

In Australia the Boab is the only member of the genus *Adansonia*, and is a close relative of the African Baobab (of which there is one species) and Madagascan Baobab

ONE STORY HAS IT THAT ALL ANIMALS WERE ASSIGNED A TREE. THE HYENA WAS GIVEN THE BOAB AND WAS SO UPSET THAT HE TURNED THE TREE ON ITS HEAD. ANOTHER LEGEND WAS THAT THE FIRST EVER BOAB WAS SO BEAUTIFUL, AND KNEW IT, THAT THE GODS PUNISHED IT BY ENSURING IT LIVED ITS LIFE UPSIDE DOWN.

(of which there are six). Some scientists believe millions of years ago, long after the break-up of the Gondwana landmass (which formed Australia, Antarctica, Africa, South America, New Zealand, India and New Caledonia), Baobab fruit floated from Madagascar and took root in Australia's north-west.

It is fitting, therefore, that the much travelled Boab—*Adansonia gregorii*—was named after the northern Australian explorer Augustus Charles Gregory. It was Gregory who opened up large parts of Western Australia, Queensland and the Northern Territory for European settlement in the mid-1800s.

Boabs thrive on creeks and drainage channels, and dot the landscape of the northern plains and tablelands, stretching from near Broome in Western Australia to north of Victoria River in the Northern Territory.

Like several trees in the Australian tropics, it sheds its leaves in the dry season. Because Boab branches look more like roots, they are shrouded in myth. One story has it that all animals were assigned a tree. The hyena was given the Boab and was so upset that he turned the tree on its head. Another legend was that the first ever Boab was so beautiful, and knew it, that the gods punished it by ensuring it lived its life upside down.

RIVER RED GUM

Eucalyptus camaldulensis

Greenough
Western Australia

IT IS HARD TO BELIEVE THIS TREE AT GREENOUGH, 400 KILO-metres north of Perth, is a River Red Gum, the same species as the proud giants that grow along river banks and in paddocks in the eastern states.

This tree is hunched over, Quasimodo-like, as if it is frightened to leave the ground that nourishes it.

There has been much debate over the years how this tree, and other similar trees nearby, came to be like this. It is now generally acknowledged that their shape is a result of the salt-laced winds that blow in from the Indian Ocean 2 kilometres away, and race over the dunes and across the Greenough river flats.

The wind-blown sea salt collects on the tree's trunk and branches, retarding their growth. The region around Greenough has a very low rainfall, which means months can pass before the salt is washed away. Trees are fantastically adaptable, and none has been more adaptable than this. The smaller its surface area, the less salt it collects, the healthier it is.

Another clue that the wind is largely responsible is that the prevailing wind is a south-westerly, which blows in each afternoon like clockwork. The tree leans to the north-east.

The land around Greenough, south of Geraldton, was cleared for pasture many years ago. Before the land was cleared, the River Red Gums would have been straighter and stronger, receiving protection from the other trees around them. River Red Gums are native in all Australian states except Tasmania. In Western Australia they are more prevalent the further north you travel. They don't venture much further south than Jurien Bay, 200 kilometres north of Perth.

The fact that this tree, most likely more than 100 years old, has survived at all means it must have some salt tolerance, but life for it is clearly hard work. Growth is neither vigorous, nor vital. River Red Gums perched on the banks of the Murray or Darling rivers lead far easier lives.

THEIR SHAPE IS A RESULT OF THE SALT-LACED WINDS THAT BLOW IN FROM THE INDIAN OCEAN 2 KILOMETRES AWAY.

ANTARCTIC BEECH

Nothofagus moorei

Springbrook National Park
Queensland

SITUATED ON A DRAMATIC ESCARPMENT IN SPRINGBROOK
National Park—on the Queensland/New South Wales
border—overlooking Mount Warning, Murwillumbah
and Byron Bay, is a tree so unusual that it is hard to
believe it is of this world.

This Antarctic Beech has a circumference of 13 metres.
Three massive, contorted trunks, covered in moss, wind
skyward. The middle trunk, hollow for the first 8 metres,
twists around itself, like a giant plait. Ferns, lichen and
various fungi have made their homes in the trunks'
small hollows.

The ring of trunks has grown from a single tree. Over
many decades the tree has undergone a process of coppicing—the dying and regrowth of new stems. This ring
of growth encircles the site where a single seed took
hold of life probably some 2000 years ago.

The Antarctic Beech is one of Australia's
remaining links with Gondwana—the
ancient continent consisting of Africa,
Australia, Antarctica, New Zealand, New
Caledonia, South America and India. In a
wetter, cooler time, about 50 million years
ago, forests of Antarctic Beech were
widespread across the continent, and provided a habitat
for animals that, for the most part, have disappeared.

About 65 million years ago Australia started to drift
north from Antarctica and its climate dried dramatically.
The result was that the rainforests on Australia's east
coast retreated to isolated areas of high rainfall. The area
overlooking the escarpment in today's Springbrook
National Park was one such place.

Today, Antarctic Beech is a dominant species of the
high-altitude cool temperate rainforest of the north
coast and northern tablelands of New South Wales and
southern Queensland, and is found up to a height of
1550 metres. It is one of three species of *Nothofagus*
occurring in Australia. The other two are *Nothofagus
cunninghamii*, found in Victoria and Tasmania, and the
deciduous *Nothofagus gunnii*, found only in Tasmania.

IN A WETTER, COOLER TIME, ABOUT 50 MILLION YEARS AGO, FORESTS OF ANTARCTIC BEECH WERE WIDESPREAD ACROSS THE CONTINENT, AND PROVIDED A HABITAT FOR ANIMALS THAT, FOR THE MOST PART, HAVE DISAPPEARED.

BRUSH BOX

Lophostemon confertus

Lamington National Park
Queensland

YOU CAN STARE AT SOME TREES FOR HOURS. THE BASE OF this 1500-year-old Brush Box, not far from Elabana Falls in the Green Mountains section of Queensland's Lamington National Park, is a riotous collection of burls, cracks, crannies, ledges and fantastically bulbous protrusions.

The Lamington National Park is a 20 000-hectare world heritage rainforest 110 kilometres south of Brisbane on the New South Wales border. It boasts high mountains and dramatic escarpments, plunging waterfalls, dramatic caves, wildflowers, bubbling creeks and forests of massive trees.

And no trees are more interesting than the half dozen Brush Box giants on the switchback Box Forest walking track, 20 minutes' walk from O'Reilly's guesthouse. They have been carbon dated at 1500 years of age, making them the oldest carbon-dated trees on the Australian mainland.

Brush Box, which resemble eucalypts, are a common street tree in Melbourne (a frightening thought if they are to reach their full potential). They are an echo of an earlier climatic time. They are normally found on fringes between rainforest and open forest, the transitional zone. This tree demonstrates that the forest was once a dry/wet sclerophyll forest. Brush Box need an ash bed for their seeds to germinate and their seedlings are shade intolerant. The inference is that the rainforest arrived after the tree germinated.

The tree's two trunks have both been victims of the high winds that occasionally blast up the valley. Despite its age and its troubles, the tree has not given up; new growth sprouts from its upper branches. It is also a generous host; its rough, fibrous bark is an ideal lodging for mosses, ferns and epiphytic plants such as the Box Tree Orchid.

Most interesting, however, is the tree's hallucinogenic base. Stand back and let your eyes wander and, with a dose of imagination, you will see contorted and grotesque apparitions and ghoulish visages—a crotchety old grandmother, a yawning ghost, a wart-covered goblin, a wizened dwarf and a cackling devil. The tree's roots, which tumble down the hill towards the path, envelop a massive rock.

A large platform several metres uphill from the rock makes a fine writer's eyrie. Sit in it and you feel like you are part of this tree, enveloped in its warm embrace.

This tree has a huge soul.

THEY HAVE BEEN CARBON DATED AT 1500 YEARS OF AGE, MAKING THEM THE OLDEST CARBON-DATED TREES ON THE AUSTRALIAN MAINLAND.

GREEN FIG

Ficus virens

Atherton Tablelands
Queensland

WHICH TREE IS AUSTRALIA'S BROADEST? THERE MAY BE NONE to beat this extraordinary Green Fig on Boar Pocket Road, an hour's drive south-west of Cairns in North Queensland.

The tree has a girth of 44 metres.

Of the world's 750-odd species of fig, most start in the ground and grow upwards. Many others, of which this is one, start their lives high in the rainforest canopy. A bird or bat drops a fig seed into a branch or crevice of a tree, which then germinates. The seedling grows happily in the tree's canopy, getting nutrients from rainfall and leaf litter.

Then things turn ugly. As the fig seedling grows larger, it sends out long cable-like roots that grow down to the ground. Other roots encircle the trunk and the seedling's leafy canopy begins to spread, taking valuable light from the host tree. The fig's roots rob the host tree of nutrients and strangle it to death—not a particularly nice way to thank it for its earlier largesse. Such figs are, for obvious reasons, called strangler figs. And they can grow extraordinarily large.

For this tree, that process probably occurred around 500 years ago. Today, the host tree is no more, but the fig is thriving in the rich red basalt soil, which receives about 1500 millimetres of rain a year. Surrounding it are other healthy rainforest trees: Brush Mahogany, Black Bean, Candlenut, Red Cedar, Red Tulip Oak and Queensland Maple.

This tree's 'trunk' is made up of thousands of aerial roots, some of them 2 metres across, others thinner than a finger. Deep among the roots—in the middle of the tree—are deep labyrinths, switchback tunnels and vast antechambers big enough to stand in. The tree resembles a mythical cathedral, with spires, cloisters and flying buttresses. The crown of the tree extends over 2000 square metres—almost the same area as two Olympic-size swimming pools.

Millions of creatures probably live in this tree, among its tangled roots and in its broad canopy. The red globular fruit provide plentiful food for birds, insects, bats and other mammals. The tree lives with a constant cacophony of bird noises.

DEEP AMONG THE ROOTS—IN THE MIDDLE OF THE TREE—ARE DEEP LABYRINTHS, SWITCHBACK TUNNELS AND VAST ANTECHAMBERS BIG ENOUGH TO STAND IN. THE CROWN OF THE TREE EXTENDS OVER 2000 SQUARE METRES—ALMOST THE SAME AREA AS TWO OLYMPIC-SIZE SWIMMING POOLS.

But what goes around comes around. The fig has become so big that other plants have taken up location in the canopy. The parasite has become the host, the attacker the attacked.

Ten different types of epiphytes, plants that grow on other plants without deriving food from their hosts, live among its trunks and branches, including Basket Ferns, Bird's Nest Ferns, Elkhorn Ferns and Staghorn Ferns. Figs are, for want of a better word, suckers for epiphytes because they do not shed their bark, and their long flat branches allow such plants to gain a secure footing.

This tree will eventually host so many other plants, animals and insects that it will begin to deteriorate. Also fires will cause damage, allowing fungi to take hold and cause decay. Termites and other insects and animals will then progressively inhabit the tree, further reducing its vigour and health. Eventually, the tree's roots and stems will become riddled with pockets of decay. Large limbs will become too heavy and fall off, causing an uneven weight distribution and destabilisation. A high wind will eventually administer the last rites.

But what a life from one tiny seed that once lodged high in another tree's branches.

WESTERN MYALL

Acacia papyrocarpa

Commonwealth Hill Station
South Australia

IN APRIL 1970, SCIENCE STUDENT BRENDAN LAY WAS traipsing through the desert 500 kilometres north-west of Port Augusta in outback South Australia. He stopped in his tracks when, among the mulga trees and saltbush eking a living from the harsh red earth, he spied an exceptionally large and spreading Western Myall, far larger than any he had seen before.

Not only was its shape remarkable but Lay was perplexed how such a tree had grown, and survived, miles away from any other Western Myalls.

The tree sits in the north-west corner of *Commonwealth Hill Station*, a 1.2 million hectare sheep station 250 kilometres south-west of Coober Pedy and 900 kilometres from Adelaide. It is only 8 kilometres from the dingo-proof fence that runs from Penong to Sydney.

The tree is not tall, about 5 metres at its highest point, but it is undoubtedly old—possibly 600 years. The further north you find Western Myalls, the slower they grow and the longer they live. And there are few Western Myalls north of *Commonwealth Hill Station*. It is one of the few acacia species to live to a great age.

Quite how it has survived for so long out here—where average annual rainfall is less than 200 millimetres and each year there are more than 50 days where the temperature tops 40 degrees—is anybody's guess. Although it is grazing country it is marginal land at best, and the past ten years have had below-average rainfall.

Little wonder the Department of Defence chose this area to test rockets from nearby Woomera. The air is clear so rockets can be monitored easily, and when they fall to Earth there are not many things they will bother. Only a few hardy folk live out here—principally sheep farmers and miners—together with sheep, feral camels and the odd dingo.

This tree's remarkable features are its circumference— 4.5 metres—and its form. Eight major branches, each of which would constitute a significant tree on its own, fan out from the base, some 15 metres long.

For two or three centuries the tree was just like any other Western Myall. Eventually, though, it grew so big that its branches collapsed. They buried themselves in the ground and continued growing, doubling back on each other like serpents wrestling in the red dirt. Life for this tree has not been particularly easy. Its twisted branches look tortured by the wind and heat.

Western Myalls extend from the Flinders Ranges in South Australia to Balladonia in Western Australia. They are the dominant tree in some parts of South Australia, such as Whyalla and Port Augusta, and grow to 6 metres or so in height. Locally they are prolific near the towns of Glendambo and Tarcoola, extending to the western fringes of the Nullarbor Plain.

They are generally happy in dry, limey soil, and their leaves are attractive to sheep, so their lower branches are often neatly trimmed. Their dark brown timber is very dense, and hard, and is used for ornamental woodwork. Farmers love their branches for fence posts. Finding straight ones is the challenge.

This tree is lucky to be here. In 1974 huge fires swept through this region from the south and burnt almost everything to the Northern Territory border. Western Myalls do not recover from fire, so this tree must have escaped the worst of the flames, probably aided and abetted by a colony of rabbits that lives among its roots and keeps the surrounding grass to a minimum. As Brendan Lay says: 'You can't thank bunnies for much, but they probably saved this tree.'

FOR TWO OR THREE CENTURIES THE TREE WAS JUST LIKE ANY OTHER WESTERN MYALL. EVENTUALLY, THOUGH, IT GREW SO BIG THAT ITS BRANCHES COLLAPSED. THEY BURIED THEMSELVES IN THE GROUND AND CONTINUED GROWING, DOUBLING BACK ON EACH OTHER LIKE SERPENTS WRESTLING IN THE RED DIRT.

HUON PINE
Lagarostrobos franklinii

Mount Read
Western Tasmania

SINCE THE END OF THE LAST ICE AGE, ABOUT 10 000 YEARS AGO, something very strange has been happening on the steep, rugged slopes of a mountain in Tasmania's remote west. An hour's drive from the small mining town of Rosebery north of Queenstown—at the end of a bouncing and winding dirt road on Mount Read—is a remote and dogged Huon Pine that has been regenerating for more than ten millennia.

After much study, scientists have proved that the several hundred Huon Pines in the region—covering about one hectare—are all male and genetically identical, which means that the pines are all clones of a single, original tree.

Importantly, there are no other Huon Pines around (the nearest other stand is at Lake Beatrice in the West Coast Range, some 20 kilometres away), and no female trees in the area. Pollen from the Huon Pines was discovered in the sediments of the nearby glacial Lake Johnston, and carbon-dated at 10 700 years.

It is most likely that the tree was part of a small forest which established itself in the last interglacial period—between 15 000 and 20 000 years ago—when conditions were warmer. The tree was the only one to survive. Since then the main method of growth seems to have occurred during harsh winters, when the tree's branches were weighed down by snow. They made contact with the ground, took root, and have continued growing.

This process has been going on for thousands of years. The tree has produced many new trunks, eventually producing a small forest of identical and interconnected trees covering the best part of a hectare. This layering method of growth is common in many plants, and in some other Tasmanian pines, such as the Pencil Pine.

This is an amazing story of dogged perseverance and longevity, especially given that the trees are at an altitude of 1000 metres, far higher than any other Huon Pines in the state.

The Mount Read area is part of a mineral lease and it is eerie in the extreme, more often than not covered in cloud or swirling with snow. Scattered throughout the forest are the haunting burnt-out remains of King Billy Pines (*Athrotaxis selaginoides*).

IT IS MOST LIKELY THAT THE TREE WAS PART OF A SMALL FOREST WHICH ESTABLISHED ITSELF IN THE LAST INTERGLACIAL PERIOD— BETWEEN 15 000 AND 20 000 YEARS AGO—WHEN CONDITIONS WERE WARMER.

The area boasts a selection of other trees, some of which could be more than 1000 years old, including many of Tasmania's other endemic conifers: Pencil Pine (*Athrotaxis cupressoides*), Cheshunt Pine (*Diselma archeri*), Creeping Pine (*Microcachrys tetragona*), and Celery-top Pine (*Phyllocladus aspleniifolius*). There are also the rare endemic Toothed Orites (*Orites milliganii*) and Tanglefoot Beech (*Nothofagus gunnii*), Tasmania's only native deciduous tree.

Huon Pines grow in forests along the rivers of western Tasmania and along the Huon River in the south, where the tree was first discovered. This was not long after Hobart Town was established in 1804. The tree was extensively logged throughout the 1800s until the mid-1900s.

Huon Pines are known to live for about 3000 years, which makes the species one of the longest living in the world. The world's oldest known individual tree is a gnarled Bristlecone Pine (*Pinus longaeva*) on a mountain top in California. Scientists bored its trunk and revealed it to be nearly 5000 years old. The giant Alerce of Chile (*Fitzroya cupressoides*) is thought to live to 3500 years.

Huon Pines grow very slowly, so they have a dense grain and narrow growth rings. For this reason, Huon Pine timber has long been favoured for furniture and boats. Its durability is due to a unique oil—methyl eugenol—that permeates the wood and makes it resistant to rot and borers, as well as giving the timber a distinctive sweet smell. Many Huon Pine logs have been retrieved from under metres of mud and sand in the Stanley River and are still as new, despite carbon dating showing they are more than 30 000 years old.

WOLLEMI PINE

Wollemia nobilis

Wollemi National Park
New South Wales

ON 10 DECEMBER 1994, ON A WEEKEND OFF, AVID BUSHWALKER and National Parks officer David Noble abseiled off a cliff and over a waterfall in the vast and rugged Wollemi National Park, 150 kilometres north-west of Sydney. He found himself in a very deep canyon surrounded by trees the like of which he had never seen before. These trees had leaves that were serrated like a dinosaur's tail and their bark was curiously bubbly.

He took a small sample of foliage back to Sydney and presented it to his colleagues at the New South Wales National Parks and Wildlife Service. Botanist Wyn Jones suggested it was a piece of fern. That might have been the end of the matter but for Noble's reply, 'It's not a fern, it's a bloody tree!'

A small team headed by Jones, his colleague Jan Allen, and Ken Hill—a conifer expert at Sydney's Royal Botanic Gardens—visited the site and realised that the tree was a new species, related to the Monkey Puzzle and Norfolk Island Pine. Before long the tree was christened the Wollemi Pine. It was declared a new genus, its scientific name, *Wollemia nobilis*, recognising the location of the find and its discoverer, David Noble.

To say that botanists were amazed at the discovery is an understatement. It made news around the world. The Wollemi Pine, it was announced breathlessly, was one of the greatest living fossils discovered in the twentieth century, with its roots back in the early Cretaceous period—110–120 million years ago—when dinosaurs roamed the Earth. Central to its survival may have been its ability to produce new tree trunks from the base (known as coppicing), allowing individual trees to regenerate when other competing species around it died.

Professor Carrick Chambers, director of Sydney's Royal Botanic Gardens, announced that it was the 'equivalent of finding a small dinosaur alive on Earth'. Sir David Attenborough said, 'How marvellous and exciting that we should have discovered this rare survivor from such an ancient past.'

Dr Tim Entwisle, executive director of Sydney's Botanic Gardens Trust, said he went through the classic stages of botanical shock: disbelief, amazement and excitement. 'We only knew of this fascinating tree from fossil fragments scattered across Australia,' he wrote. 'The Wollemi Pine is a unique reminder that the world

THE WOLLEMI PINE WAS ONE OF THE GREATEST LIVING FOSSILS DISCOVERED IN THE TWENTIETH CENTURY, WITH ITS ROOTS BACK IN THE EARLY CRETACEOUS PERIOD— 110–120 MILLION YEARS AGO—WHEN DINOSAURS ROAMED THE EARTH.

EXPERTS WERE PERPLEXED HOW
THE WOLLEMI PINE HAD MANAGED
TO AVOID TOTAL EXTINCTION,
LIVING THROUGH RAGING
BUSHFIRES AND NO LESS THAN
SEVENTEEN ICE AGES.

is full of undiscovered wonders, that there is a lot more to know about our planet and a lot to protect.'

The canyon in which the trees were found has been kept a secret to protect the survivors. After much exploration by senior National Parks staff, only 100 other adult trees were discovered in three or four patches within the complex gorge system. Experts were perplexed how the Wollemi Pine had managed to avoid total extinction, living through raging bushfires and no less than seventeen ice ages. Scientists believe that, as the continent dried out over several thousand years, the trees retreated to a single canyon in the national park where the species miraculously survived.

After the worldwide excitement at the Wollemi Pine's discovery, there was bound to be great interest when the first trees cultivated from the rare conifers were made available to the public. A special auction was held at the Sydney Botanic Gardens in 2005 attended by the world's press. Lot 1, a single tree propagated from the Bill Tree, went under the hammer for $6000. One keen collector paid $135 625 for the Sir Joseph Banks Collection of fifteen pines (one from each of the fifteen selected wild trees). The auction realised $1 059 165—exceeding wildest expectations. And all this without any American bids, due to the US embargo on importing trees taller than half a metre.

The Bill Tree is the grandest Wollemi Pine in the Wollemi National Park, and is therefore the grandest Wollemi Pine on Earth. Nearly 40 metres tall, it rises high above the canopy, and is judged to be 1000 years old. It is named after Bill Hollingsworth, the American-born helicopter crewman responsible for the safe transport of the scientific team on their early expeditions. Author James Woodford describes the Bill Tree as 'a reclusive biological superstar; a fossil turned green before science's eyes'.

BANYAN FIG

Ficus benghalensis

Brisbane Botanic Gardens
Queensland

STROLL PAST THE HISTORIC QUEENSLAND CLUB IN CENTRAL Brisbane and head south-east towards the Brisbane River, leaving Parliament House on your right. Enter the Botanic Gardens, Queensland's oldest gardens, and soon you will be standing under a Banyan Fig that is so unusual you will not be able to stifle a chuckle.

The tree's colossal central trunk is made up of hundreds of fused roots. Up to 15 metres away from the main trunk are other subsidiary trunks, some with diameters of up to a metre. The tree's canopy measures 50 metres across. You don't stand under this tree so much as stand *in* it.

The Banyan—also known as a Bengal Fig, Indian Fig, East Indian Fig and Indian Banyan—is India's national tree and is endemic to India, Sri Lanka and Bangladesh. It is one of the world's largest tree species in terms of area, largely because of the way it grows. It puts down aerial roots, which descend like ropes from the branches and enter the ground. Then they take root and grow into subsidiary trunks. These subsidiary trunks either merge with the main trunk or continue to grow apart from it.

Banyan Figs are famous for covering extremely large areas. One tree in Kolkata (Calcutta), India, covers an area of 1.5 hectares, and has more than 2880 subsidiary trunks. The main trunk decayed and had to be removed in 1925; the remaining prop roots have flourished and the Banyan now looks more like a forest.

This Brisbane tree was planted in the 1870s, so it is just over half the age of the famous Kolkata Banyan. The longer you stand under this tree, the more you will see. The various prop roots have assumed grotesque and hideous shapes. You can see faces contorted in agony, hands clutching at air, and outlines of bodies tumbling in freefall. Dante would have loved this tree.

The Banyan Fig got its name from the Hindu merchant caste, the Banyans, who set up their stores under the shelter of these expansive trees. One Hindu belief is that a Brahmin was turned into a Banyan Fig and his spirit lives on in the tree. Because of this the trees are sacred to the Hindus and are encouraged to grow and are rarely pruned, even though Banyans can cause great damage to temples and buildings.

This species has achieved fame far and wide. It is, for example, part of the coat of arms of Indonesia. The most famous castaway of all, Robinson Crusoe, made his home in a Banyan Fig.

The Banyan is one of more than 750 species of fig, more numerous than oaks and eucalypts. Extraordinarily, each kind of fig has its own species of wasp. In each case they form a beautiful symbiotic relationship. The female wasp lays her eggs in her particular fig's fruit and the young wasps hatch out, mate, and the females then fly to another fruit to lay eggs, so dispersing the first fig's pollen. For tens of millions of years each species of wasp has evolved alongside the species of fig that it pollinates.

FOREIGN
INVADERS

GOLDEN ELM
Ulmus glabra 'Lutescens'

South Yarra
Melbourne, Victoria

IN THE MID-1850s PUNT ROAD ON THE south side of Melbourne's Yarra River was little more than a dusty track, a conduit for people heading to or from the city on horseback, in carriages, or on foot. A precarious punt ferried people across the river. On the west side of the road, towards the top of the hill, was a grand house owned by Scottish lawyer David Ogilvy which he called *Airlie Bank* after his ancestral home in Scotland. Ogilvy's grapevines stretched north from his house right down to the river bank.

By the 1880s the value of Ogilvy's land had skyrocketed and in 1886 he subdivided. Today South Yarra is one of Melbourne's densest suburbs and Punt Road is one of Australia's busiest, carrying 73 000 cars a day. But none of this seems to bother a Golden Elm living close to the noise and bustle, which many consider the city's most beautiful tree.

The elm, which grows on a small reserve on the corner of Alexandra Avenue where it joins Punt Road, was planted in 1938 after the Melbourne City Council bought the land as part of a reservation for the further widening of the road. Today the tree is such an icon it even has its own entry in the city's street directory.

The tree is a picture of health despite its harsh urban surroundings. Although it is only 12 metres high, its canopy spreads over 22 metres, and shows no sign of stopping there. Its rounded branches fill completely the small reserve where it lives.

To fully appreciate the tree, you must stand beneath its canopy. Its long branches curve gracefully outwards, the lower limbs touching the ground. It is a hidden haven, far removed spiritually from the hustle and bustle of harsh city life. The tree is clearly much loved; some of its more vulnerable branches have been propped up and others are supported by cables.

There is no better time to see the tree than in spring-time. Its leaves are a luminescent, deep gold, contrasting with the darker green of nearby trees, principally English Elms (*Ulmus procera*). It is surrounded by *Dietes*—a drought-resistant South African plant with purple and white flowers.

Australia's elms have so far escaped the ravages of Dutch Elm Disease, spread by the elm bark beetle which has killed elms around the world. However, the elm leaf beetle is a perennial threat to the Golden Elm's health, and the Melbourne City Council regularly treats the tree to keep the beetle in check.

The council arborists also notice from time to time that ashes have been scattered under the tree, presumably the result of cremations. This indicates the place the tree holds in many hearts.

ALGERIAN OAK

Quercus canariensis

Royal Botanic Gardens
Melbourne, Victoria

IN JULY 1900 ENGLISHMAN JOHN Adrian Louis Hope, the seventh Earl of Hopetoun, was appointed Australia's first Governor-General. He arrived in Sydney in December of the same year, amid much fanfare.

However, the Earl had lived in Australia before. Between 1889 (when he was only twenty-nine) and 1895, he served as Governor of Victoria, stationed at Melbourne's Government House, the large and elegant Italianate-style building overlooking Melbourne and the city's famous Botanic Gardens.

Lord Hopetoun had made the most of his time in Victoria. His chief interest was hunting on horseback, a hobby one of his vice-regal successors, Sir George Clark, said had caused him to 'break more bones than any man of his time in England'. *The Bulletin* said that the Earl had a 'willowy stoop' and his habit of wearing hair-powder raised local eyebrows. But he was popular for his unassuming and friendly nature, his love of riding (he often took informal horseback tours around town) and sports and his lavish entertaining at Government House, where he had the assistance of fifty-five servants.

In March 1895, after six eventful years in Melbourne, he and his family—his wife, the Countess of Hopetoun, and their sons Lord Hope and the Honourable Melbourne Hope—prepared to return to England. On the day of their departure each planted an Algerian Oak on the Botanic Gardens' Hopetoun Lawn, the sloping expanse between Government House and the Ornamental Lake, looking east over the suburbs of Richmond and South Yarra.

Today, 114 years later, by far the largest of these Oaks is the one planted by little Melbourne Hope, who was only four at the time. It is 24 metres high and has a canopy diameter of 37 metres. Its trunk is 4.6 metres in circumference at chest height. In 1938, in his book *In the Botanic Gardens*, author Frank Clarke wrote: 'Melbourne Hope's tree is twice the size of the others because it is grown on an old sunken path filled in.'

The tree is a special favourite with those who frequent the gardens because of its beautiful proportions and grand sweeping branches. In March 1995, when the trees turned 100, a bemused gardener witnessed four people celebrating their coming of age with ribbons, balloons, incense and chanting.

The roots of Melbourne Hope's tree disappear into the soil with purpose, epitomising strength and resilience. For much of the year the tree's large leaves filter the morning light, leaving dappled outlines on the ground. Algerian Oaks are semi-evergreen in Melbourne's

TODAY, 114 YEARS LATER, BY FAR THE LARGEST OF THE OAKS IS THE ONE PLANTED BY LITTLE MELBOURNE HOPE, WHO WAS ONLY FOUR AT THE TIME.

THE ALGERIAN OAK IS NATIVE
TO PORTUGAL, SPAIN, TUNISIA,
ALGERIA AND MOROCCO. ITS
ABILITY TO THRIVE IN DUSTY, DRY
SOIL AND HOT CONDITIONS MAKES
IT A FAVOURITE IN AUSTRALIA.

climate, losing only some of their leaves when they are pushed off by new growth. English Oaks, by contrast, are largely deciduous in this climate.

The Algerian Oak's scientific name, *Quercus canariensis*, suggests the tree is a native of the Canary Islands off Africa's west coast. This is not the case, although many have been planted there over the centuries. The tree is in fact native to Portugal, Spain, Tunisia, Algeria and Morocco.

It is this heritage—the ability to thrive in dusty, dry soil and hot conditions—that makes the Algerian Oak a favourite in Australia. The tree does not need much water and its well-leaved branches provide excellent shade. 'We will look to these sorts of trees when planting for potentially drier conditions,' says Richard Barley, the current director of the gardens.

The man most responsible for importing Algerian Oak acorns into Melbourne was Ferdinand von Mueller, director of the gardens between 1857 and 1873. Many of his acorns were planted throughout the state, and many resulting trees—such as the one planted in Kyneton in May 1863 to celebrate the marriage between Edward, Prince of Wales and Princess Alexandra of Denmark— are still alive. We do not know whether the acorns that Lord Hopetoun and his family planted in Melbourne were imported, or came from a tree already growing in Australia.

Richard Barley says von Mueller would be 'pleasantly surprised' if he were to pay a visit to the Hopetoun Lawn today. 'It's not an easy site for trees to grow on because any water tends to run off down the hill,' he says. 'Although the site is protected from westerly winds, it is open to hot northerlies. The tree has done well because it has good genes and is used to dry conditions.'

Despite its fine progress, the tree has a long way to go to match the age of the great oaks of Europe and North Africa, some of which have lived for more than half a millennium.

CORK OAK

Quercus suber

Royal Botanic Gardens
Hobart, Tasmania

IN 1996 ARBORISTS AT HOBART'S Royal Botanic Gardens noticed with concern large clumps of mushrooms growing around the trunks of some of their more celebrated trees. Since 1818, Hobart's gardens have enjoyed a prime position in Australia's southernmost capital city, occupying 14 hectares overlooking the famous Derwent River and visited by thousands every year. This was the worst scare the gardens had had. The mushrooms indicated the presence of the deadly Honey Fungus *Armillaria luteobubalina*. This fungus penetrates the roots of a tree then travels up the trunk, effectively ringbarking it and cutting off the flow of water.

The gardens' hard-working botanists were inundated with advice from locals suggesting various potions and remedies. One correspondent was adamant that donkey urine sprinkled among the mushrooms was a sure-fire antidote.

Some members of the public even sought divine inspiration against the invader. Late one morning in spring a man dressed as a wizard with a long flowing beard came into the gardens and read incantations from a leather book—held by a young accomplice dressed in similar attire—next to the trunk of the gardens' famous 150-year-old Cork Oak, one of the affected trees. The tree, the oldest Cork Oak in Australia, is much loved, and was grown from an acorn from Blenheim Palace outside London, home of the Dukes of Marlborough and Winston Churchill.

Although about 200 trees and shrubs were lost to the fungus, many were saved when the gardens' botanists laboriously removed the infected soil, aired the roots, and covered them with clean soil.

Today the Cork Oak is in good health, despite the fact that it occasionally has to deal with flocks of Black Cockatoos, which forage in its upper limbs for bugs and caterpillars.

Thirteen metres tall with a 24-metre canopy so thick it almost blocks out the sun, this tree is the second largest of its type in Australia (the largest is at Tenterfield, New South Wales). It is certainly one of the most impressive. It has a gnarled, pockmarked trunk and a fine soccer ball-sized outgrowth—a burl—protrudes a metre from the ground. It produces acorns intermittently. The tree occupies a lordly location on the main lawn, surrounded by Norfolk Island Pines, a Laurel, various melaleucas and an impressive New Zealand Christmas Bush.

Cork Oaks are native to south-western Europe and north-western Africa, and are particularly common in Spain and Portugal where their springy, spongy bark is harvested every ten years or so for cork. This does not harm the trees and a new layer of cork regrows. The Cork Oak's bark is around 10 centimetres thick, which gives the tree excellent protection against fire. In Portugal it is illegal to cut down Cork Oaks.

Cork Oaks are not particularly common in Australia, but thrive best in the cooler states of Victoria, New South Wales and Tasmania. Six thousand Cork Oaks in Canberra, planted as an experiment in 1916, have recently begun to be professionally stripped.

Even in these temperate areas the trees do not survive as long as their European counterparts. In European winters the trees grow slowly and the tightly packed growth rings strengthen the trunks. European trees in Australia grow faster, year-round, and are weaker as a result.

In January 2006 a huge north wind ripped down the Derwent River, badly damaging eighteen trees in Hobart's Botanic Gardens. But, thankfully, not the Cork Oak.

OLIVE
Olea europaea

Elizabeth Farm
Parramatta, New South Wales

IN 1816 A LARGER-THAN-LIFE Englishman, having spent the best part of three decades in the new colony of New South Wales, was on a tour around Switzerland and the south of France.

This was John Macarthur, ex-soldier, landowner and amateur botanist. He was on a mission with his sons, James and William, to source plants and to research viticulture and olive production with a view to developing these industries in the antipodean colony. The cuttings and seedlings he obtained were destined for his home, *Elizabeth Farm*, in Parramatta.

On 30 September the following year a strange sight greeted those who came to the Sydney docks to witness the arrival of the convict ship *Lord Eldon*. From glasshouses erected on deck emerged a fine selection of camellias and roses interspersed with fruit trees such as figs, oranges, plums, lemons, quinces, grapes, pomegranates, guavas and olives. Macarthur's diary notes that he had 'olives in four pots and two olive plants from Provence'.

Two ancient Olive trees are still thriving in the garden of *Elizabeth Farm*. There is a good chance that they were among the seedlings Macarthur brought out in 1817.

Then again the trees might be older still. In 1825 a French naval officer, Hyacinthe Yves Philippe Potentien, Baron de Bougainville, wrote in his journal that he visited *Elizabeth Farm* and that John Macarthur 'gave us a guided tour of his garden which he cultivated with loving care and where flourished all manner of fruit trees. We saw … a very fine 20-year-old Olive tree laden with fruit from which oil is manufactured.'

Either way, the two Olives that grow today at *Elizabeth Farm* are hugely significant. They are the oldest remaining European trees at a private residence. One of them sits at the northern edge of the *Elizabeth Farm* garden, while the more celebrated of the pair grows in the middle of the old carriage loop in front of the house.

This tree is certainly showing its age. Three years ago disease in the trunk necessitated a severe pruning. One ancient limb now lies in pieces at its base, while another continues to survive and several newer and thinner trunks are thriving.

Every second October the Festival of the Olive is held at *Elizabeth Farm*, celebrating Mediterranean culture in Australia. Up to 7000 people attend. The old Olive tree is the star attraction, as it was the first indicator that Mediterranean trees could be grown in Australia. Every three years or so the trees fruit abundantly, producing small black olives.

The house at *Elizabeth Farm* was built in 1793 and is still intact, run by the Historic Houses Trust of New South Wales. Macarthur and his wife Elizabeth lived there for many years, as their sheep empire grew. In 1832 John was declared a lunatic and died two years later. By that stage his family had amassed 9600 hectares

of land—including property at Camden and along the Southern Highlands as far as Goulburn—and held mortgages over another 5200 hectares. Elizabeth outlived her husband by sixteen years.

By any measure Macarthur, described variously as headstrong, canny, arrogant and pigheaded, lived a full life. In 1801 he was involved in a duel with his commanding officer Lieutenant-Governor William Paterson, who was injured. Macarthur was recalled to England to defend his reputation. While in England he convinced Joseph Banks, who was in charge of the King's flock, to allow him to take a small number of Spanish merino sheep back to Australia. As every Australian student knows, this was the start of the country's most famous industry.

A few years later, in 1808, he was involved in the military overthrow of Governor Bligh, an event known as the Rum Rebellion. Again Macarthur returned to England, to advise a colleague, Captain George Johnston, during his court martial. Macarthur would not return to Australia for eight years. However, he spent much of his time in England marketing Australian wool.

The garden at *Elizabeth Farm* contains other introduced trees besides the aged Olives. Species include two Cherry Laurels (*Prunus laurocerasus*) that were described as 'flowering profusely' when Elizabeth Macarthur was alive. However, in the last decade they have had only two flowerings. Among the exotics are a Chinese Elm (*Ulmus parvifolia*), an aged Japanese Pagoda Tree (*Sophora japonica*) with its replacement seedling growing nearby, as well as a California Fan Palm (*Washingtonia filifera*). There are also Hoop Pines, Bunya Pines and several fine cacti.

TWO ANCIENT OLIVE TREES ARE STILL THRIVING IN THE GARDEN OF *ELIZABETH FARM*. THERE IS A GOOD CHANCE THAT THEY WERE AMONG THE SEEDLINGS MACARTHUR BROUGHT OUT IN 1817.

HISTORIC TREES

BOAB
Adansonia gregorii

Prince Regent Nature Reserve
North-west Kimberley, Western Australia

ON 21 SEPTEMBER 1820—NINE years before the city of Perth was founded—His Majesty's cutter *Mermaid*, under the command of the steely Phillip Parker King, was careened at a small scimitar-shaped bay. Rather unimaginatively named Careening Bay, it was a perfect spot for their ship's overhaul, protected by two headlands of basalt rock on the north-west coast of the remote Kimberley region of what is now Western Australia.

There, on the gently sloping beach, King and his crew carried out much-needed repairs to their boat before continuing their voyage. King had been given instructions by the British Secretary of State for the Colonies to: 'Take care to leave some evidence which cannot be mistaken of your having landed.' He carved the name of his boat, and the year of their visit, into a two-trunked Boab growing 100 metres from the water mid-way along the beach.

On his return to Careening Bay the following year in the brig *Bathhurst*, King noted, 'The large gouty-stemmed tree on which the *Mermaid*'s name had been carved in deep indented characters remained without any alteration, and seemed likely to bear the marks of our visit longer than any other memento we had left.' Could this Boab, which still clearly shows the name of King's boat in big bold letters, carry Australia's oldest graffiti?

This stately tree, which lies between Port Nelson and Cape Brewster in the Bonaparte Archipelago 300 kilometres north-east of Derby, is still healthy. It is sustained, no doubt, by a small stream that flows down from the hills behind it.

The government has recognised the importance of the region and has allocated it Nature Reserve status. Today the Prince Regent Nature Reserve is known worldwide as a UNESCO World Biosphere area, bisected by the tumbling Prince Regent River that flows off the Gardner Plateau to the south-east. King could not have found a more isolated area to land his little boat. Even today it is so remote that only the most determined traveller, who can reach the spot solely by air or sea, will be able to visit this stout marker of Australia's early history.

Its divided trunk, a common characteristic of Boabs, supports a range of elephantine-shaped bulbous branches. The tree is covered in creepers which die back in the dry winter months. It has a 12.5-metre girth and appears very content here, showing signs of neither age nor distress.

In 1992, the crew of HMAS *Cessnock* paid a visit to Careening Bay and some of them carved the name of their ship into the bark of a Boab 200 metres south of the Mermaid tree. The navy hauled the graffitists over the coals for their efforts.

How things have changed.

RIVER RED GUM

Eucalyptus camaldulensis

Wilpena Pound
Flinders Ranges, South Australia

IN MAY 1937 CELEBRATED Australian photographer Harold Cazneaux (with his wife Winifred and their son Harold) drove north from Adelaide through the South Australian towns of Port Pirie, Quorn and Hawker. Their destination was Wilpena Pound, the rock amphitheatre in the heart of the prehistoric Flinders Ranges, 450 kilometres north of Adelaide.

For several years Cazneaux had been fascinated by the saw-toothed ranges that punctuated the harsh, arid countryside to the north of South Australia's famous Clare Valley. He was particularly drawn to the light, the earthy colours, and the drama of the landscape.

During his week-long stay in the ranges, Cazneaux was captivated by an ancient River Red Gum, 4 kilometres east of Wilpena Pound. He took several photos of the tree and, in 1941, nominated one of these photos as his 'most Australian picture'. The photo, which he titled 'The Spirit of Endurance', was published in *Australia: National Journal* alongside the following words:

This giant gum tree stands in solitary grandeur on a lonely plateau in the arid Flinders ranges, near Central Australia—where it has grown up from a sapling through the years, and long before the shade cast from its giant limbs ever gave shelter from heat to white man. The passing of the years has left it scarred and marked by the elements—storm, fire, water—unconquered, it speaks to us of a spirit of endurance. Although aged, its widespread limbs speak of a vitality that will carry on for many more years. One day, when the sun shone hot and strong, I stood before this giant in silent wonder and admiration. The hot wind stirred its leafy boughs and some of the living element of this tree passed to me in understanding and friendliness expressing the Spirit of Australia.

Cazneaux died in Sydney in 1953. He would doubtless be pleased that his tree is still alive, if a little weather-beaten and weary, more than seventy years after he took his famous photo.

The last ten years of drought have not helped this statuesque gum, but it clings to life, and the dry soil that supports it, with tenacity. Its roots, exposed over the centuries by the wind and the water from the creek on which it lives, are bleached and battle-hardened, resembling bony fingers clinging to the dusty earth.

The inside of the tree, just as it was when Cazneaux photographed it, is completely hollowed out, probably burnt by Aboriginal people many decades ago. Where the bark has peeled off, the lower section of the trunk gleams in the fading evening light, the jagged eastern slopes of Wilpena Pound providing a startling backdrop.

The tree's base is littered with fallen branches, some considerable, but there is new life too—green leaves sprout from the lower part of the trunk, as well as from the tree's crown.

Despite this ongoing tilt at life, you get the impression that Cazneaux's tree is not long for this world. This battle-scarred symbol of fortitude and resilience, recognised and captured by one of Australia's foremost photographers, will soon return to the earth from which it emerged. But Cazneaux's famous photograph will ensure its memory lives on.

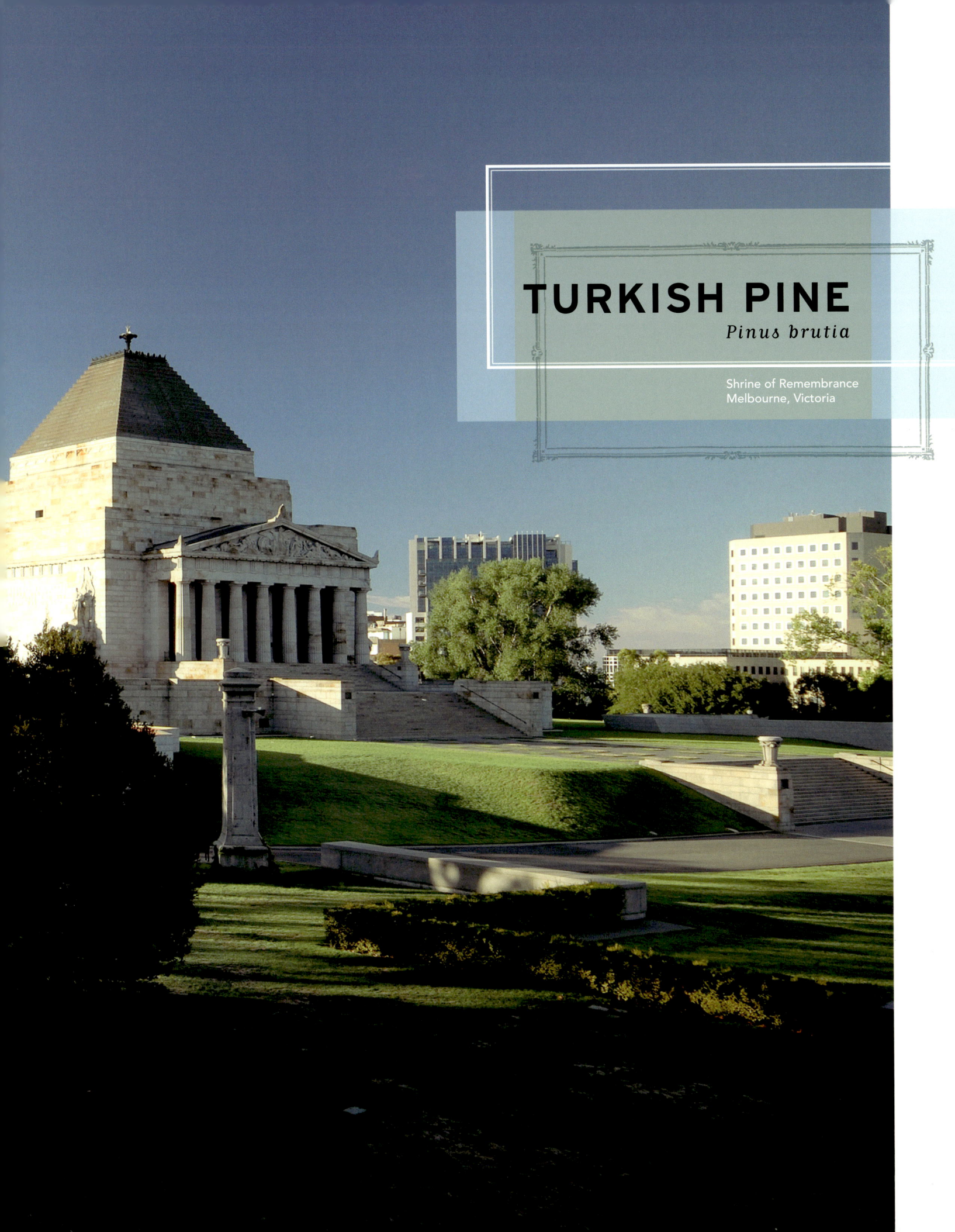

TURKISH PINE
Pinus brutia

Shrine of Remembrance
Melbourne, Victoria

ON 6 AUGUST 1915 A SINGLE
Turkish Pine high on a plateau
on the Gallipoli Peninsula in
Turkey witnessed some of the
fiercest fighting of World War I.
Australian and New Zealand
troops attempted to gain control of Turkish trenches
under heavy machine-gun and artillery fire. They
achieved their purpose at great cost, got access to the
trenches and engaged in brutal hand-to-hand combat.

The fighting continued for four days and the six
Australian battalions involved lost 2197 men and eighty
officers. Between 5000 and 6000 Turks were killed.
There was much heroism on both sides and seven
Victoria Crosses were awarded to Australian soldiers.

The pine on the plateau (*Pinus brutia*, also known as
a Calabrian Pine, East Mediterranean Pine and Brutia
Pine) was destroyed in the fighting but a soldier,
Sergeant Keith McDowell of the 24th Battalion,
managed to salvage a cone from the fallen tree. He
carried it with him in his haversack, as a memento,
until the end of the war.

McDowell returned to Australia and gave the cone
to his aunt, Emma Gray, who lived at Grassmere, near
Warrnambool in Victoria. A decade later she planted
the seeds and grew four seedlings.

The significance of these seedlings was profound. On
11 June 1933 Lieutenant-General Sir Stanley Savige
planted one of the seedlings, with full military honours,
at the Shrine of Remembrance in Melbourne.

Today the 80-year-old tree stands, minus a few pruned
limbs, near other military reminders—the Eternal Flame
and a bronze statue of Simpson and his Donkey. It is a
reminder of courage and valour in the face of extreme odds.

Seedlings have been grown from Melbourne's Lone
Pine under the direction of Melbourne Legacy's
Commemoration Committee, which is responsible for
the collection, propagation, presentation and dedication
of pines from the tree at the Shrine.

Every year on Anzac Day, before dawn breaks, thousands
of people gather under the tree and around the Eternal
Flame to pay their respects to Australia's war heroes.

TODAY THE 80-YEAR-OLD TREE
STANDS AS A REMINDER OF COURAGE
AND VALOUR IN THE FACE OF
EXTREME ODDS.

RIVER RED GUM

Eucalyptus camaldulensis

Springton
South Australia

ON THE OUTSKIRTS OF THE SMALL TOWN OF SPRINGTON IN South Australia's Adelaide Hills is an unusual tree, with a scarcely believable history.

One and a half centuries ago—on 6 June 1855—the barque *Wilhelmine* left Bremen, Germany's oldest port city, for the 26 000-kilometre journey to Australia via the west coast of Africa, the Cape of Good Hope and the Indian Ocean.

Among the 110 passengers was 27-year-old tailor Johann Friedrich Herbig. He was eager to leave Europe behind, where the Crimean War was raging, and start life again in the new colony of South Australia. One hundred and twenty-five days later the passengers sighted the western tip of Kangaroo Island, and the journey was over.

It had not been a pleasant one. Not long after landing the passengers signed a petition and gave it to the Governor of South Australia complaining about the ship's food. Among their gripes were peas full of maggots, mouldy dumplings, black bread, tarry water and beef 'so offensive to the smell as to be absolutely inedible'. There were fifty-five signatories to the petition, the first of which was Friedrich Herbig. By the time the authorities tried to bring the ship's captain, Mr Rimme, to justice, he had sailed for Hong Kong. The *Wilhelmine* would not return to Australia for twenty-six years.

In 1855 the population of South Australia was approaching 100 000. The streets of Adelaide were full of horse-drawn carts, dogs and scampering children. Food was not expensive; the London Soup and Chop House served kangaroo tail soup with sherry for one shilling a serve, and the finest brandy cost fourteen shillings a gallon. All manner of businesses had sprung up including Mr Phillips the dentist, who specialised in non-corrodible mineral teeth, 'from one to a complete set'.

Friedrich Herbig turned his back on this city of possibility and headed north to Blumberg (now Birdwood), a German settlement in the Murray Ranges. He had heard that employment was on offer at Black Springs (now Springton), a further 20 kilometres to the north, where English philanthropist George Fife Angas had started a dairy.

But where to live? At Black Springs Herbig spotted an old gum tree with a hollowed-out middle. The tree, he reasoned, seemed as good a place as any to set up home. So he did.

The tree was broad and squat, its base measuring 6 metres across and its height around 25 metres. Many years earlier the tree had lost its upper trunk and its inside had been burnt out by fire. Most importantly for Herbig the tree's opening—or, in his case, the doorway—faced east, away from the prevailing wind and rain.

Herbig worked at the dairy and leased 80 acres of land from his employer, which included his tree house. Eleven years later he exercised his option and bought the land.

During the summer of 1856, the brig *Vesta* arrived in Port Adelaide. Among those disembarking were the

AT BLACK SPRINGS HERBIG SPOTTED AN OLD GUM TREE WITH A HOLLOWED-OUT MIDDLE. THE TREE, HE REASONED, SEEMED AS GOOD A PLACE AS ANY TO SET UP HOME. SO HE DID.

Rattey brothers, Erdmann and Gottfried, and their wives and families. With them came their 16-year-old niece Anna Caroline Rattey. All were illiterate German-speaking Polish peasant farmers of Russian extraction looking for a new life in the colony.

The next year Caroline, as she was known, had a frightening experience at Cockatoo Valley on the edge of the Barossa Valley. A rogue stabbed her in the chest and strung her up in a wattle tree, leaving her to die. The tree bent enough for her feet to touch the ground. She managed to escape and ran half a mile to a farmhouse to get help.

Unnerved by her experience, Caroline made her way to Black Springs, to live with one of her uncles. There she met Friedrich—the strange chap who lived in a tree—and in July 1858 they got married. She moved in with him.

History doesn't relate whether Caroline was greatly perturbed by the fact that their house was not a conventional one. A year later the first of their sixteen children was born. In September 1860 another son arrived, and the couple realised that they had better seek more extensive accommodation. They built a two-room hut nearby with solid red gum beams and a thatched roof.

Friedrich Herbig became a chaff merchant, ably assisted by his growing brood, and by the time he died in 1886 (falling from a wagon while crossing a creek) he owned 1000 acres of land. He and Caroline, who died in 1927—outliving her husband by 40 years—have 860 descendants. There are still four families in Springton who claim Friedrich and Caroline Herbig as their forebears.

And what of their old tree? Well, it is certainly alive and is a local Springton attraction. Descendant David Herbig, who lives nearby, tends it carefully and says it looks healthier today than it did forty years ago. He attributes this to the affection it receives from the many Herbig descendants who visit regularly and wonder at Friedrich and Caroline's resilience.

The tree, probably more than 500 years old, exceeds 16 metres in circumference, which puts it among the broadest River Red Gums in the land. A collection of burls makes up its gnarly base. Overhead, dead limbs intermingle with new limbs, tinged with bright green leaves.

No tree in Australia tells a better story about the challenges that had to be overcome by Australia's early immigrants, nor better underlines the single-mindedness that many of them possessed.

COOLIBAH
Eucalyptus coolabah

Cooper Creek
Near Innamincka, south-west Queensland

WHEN WILLIAM BRAHE CARVED THE WORDS 'DIG 3FT NW' on a nondescript Coolibah tree on Cooper Creek, in what is now south-west Queensland, on 21 April 1861 before riding with a heavy heart south towards Menindee on the Darling River, he could not have imagined that the tree would end up being Australia's most iconic living symbol.

It has entered Australian folklore that, only a few hours later, three skeletal explorers—Robert O'Hara Burke, William John Wills and John King—arrived at the creek and were devastated to find that Brahe and his colleagues had left.

The explorers had travelled the best part of 4500 kilometres through Australia's harsh interior, enduring terrible hardship, and when they neared the creek they thought their troubles were at an end. However, the camp was silent and the stockade, which had housed Brahe and his team for four months, was empty. (In fact Brahe had already stayed a month longer than Burke had ordered.)

The explorers saw Brahe's blaze on the tree and dug up the cache of provisions. They set out for Mount Hopeless in South Australia but knew their attempt was useless and so returned to Cooper Creek. Burke and Wills later died on the river bank, some kilometres downstream from the tree. King was rescued and lived to tell the remarkable tale of the expedition's ill-fated trip from Melbourne to the Gulf of Carpentaria.

Today, 148 years later, 'the Dig Tree'—as the Coolibah has become universally known—is in surprisingly good health. Situated on Bullah Bullah Waterhole, it has survived floods, drought, erosion and termites (cement has been poured into the lower section of the trunk to prevent the insects spreading). A walkway has been installed around its roots to stop the surrounding soil from compacting under visitors' feet.

Brahe's blaze has largely been grown over but Burke's earlier mark facing the creek, consisting of the letter B for Burke and the camp number LXV (camp 65) underneath, is still clearly visible.

The tree, probably close to 350 years old, sits some way up the bank, its long roots spreading southwards to the river's life-giving waters. Brahe's stockade has disappeared, the victim of a flood some years ago. Although the river can sometimes flood tremendously, drawing its waters from the Thomson and Barcoo rivers to drain into Lake Eyre, its water mostly lies in quiet billabongs, occupied by Pelicans, Cockatoos and Corellas.

'The Dig Tree' sits on Nappa Merrie cattle station, 40 kilometres east of Innamincka, and receives its fair share of visitors, despite the fact that it is about as far away from Australia's population hubs as it is possible to get—1300 kilometres from both Sydney and Brisbane and 850 kilometres from Adelaide. Those who make the trip to the tree, and wander the eerie banks of Cooper Creek, cannot remain untouched by the location, and the grim memories it holds.

With luck, Cooper Creek will continue to nourish this tree for decades to come, honouring the deaths of those intrepid explorers, and keeping alive the memory of their monumental, and disastrous, expedition.

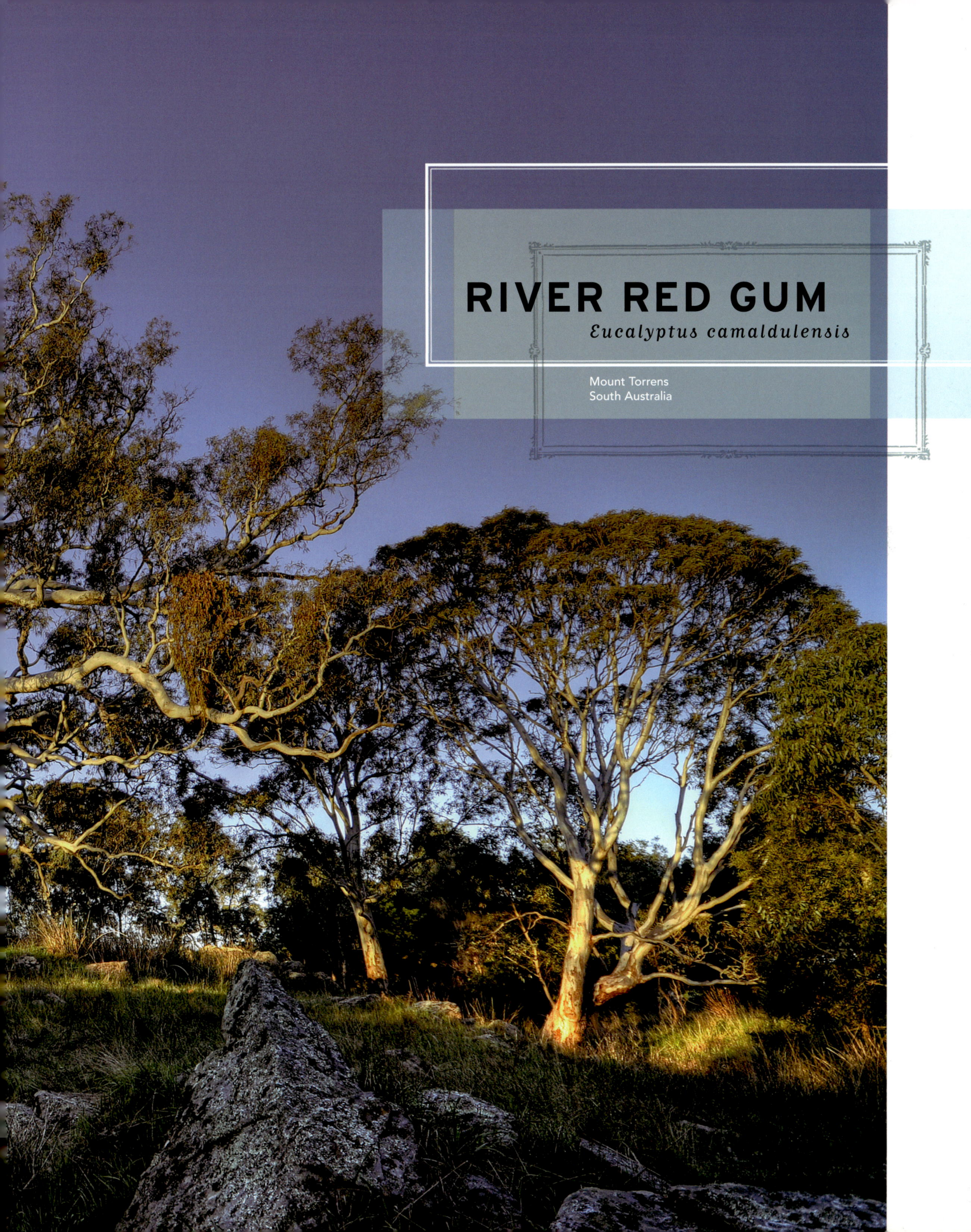

RIVER RED GUM

Eucalyptus camaldulensis

Mount Torrens
South Australia

RIVER RED GUMS ARE RARELY uniform or symmetrical. Their trunks are more often than not twisted and gnarled and their branches, like Medusa's hair, sprout out every which way. Perhaps this is why they occupy a special place in the hearts of many Australians, reflecting the unpredictability and often chaotic nature of the Australian bush.

River Red Gums are the most widespread and probably the best known of Australia's gum trees. They occur in all states except Tasmania. They can be found situated proudly on their own in the middle of a paddock, or with many others lining rivers and creeks.

Look closely at this River Red Gum, on *Murray View* station east of Mount Torrens in the Adelaide Hills, and you will see something stranger still. Several large branches splay out from the trunk and then merge together, before separating again.

Centuries ago, Aboriginal people used to weave together the branches of young trees in order to mark places of significance. As the trees grew, the branches then fused together to form rings. So-called 'ring trees' are often found on waterways and were significant cultural markers for Aboriginal communities. Some ring trees served as boundaries between lands, where strangers had to stop and seek permission to enter.

Other ring trees were markers for places of abundance. Often Aboriginal people would travel many hundreds of kilometres to meet together at certain times, such as when bogong moths descended on the Victorian and New South Wales high plains, or when the Queensland Bunya Pines were in fruit.

Long before Europeans came to the Adelaide Hills in the 1830s, the Peramangk Aboriginal people lived here, feasting on plants, grubs and animals such as kangaroos and possums, all of which were in large supply in the region's steep gullies and tablelands. They had definite borders against the Kaurna tribe of the Adelaide plains to the west and the Murray River tribes to the east. The Peramangk traded regularly with the Murray River Aboriginal people.

CENTURIES AGO, ABORIGINAL PEOPLE USED TO WEAVE TOGETHER THE BRANCHES OF YOUNG TREES IN ORDER TO MARK PLACES OF SIGNIFICANCE.

Elderly locals remember the Peramangk people and the Kaurna people meeting regularly near this tree for celebrations and corroborees. They may have honoured their dead under this ancient River Red Gum; several flat lichen-covered stones have been upturned and are pointing to the sky, possibly indicating this is a burial tree.

This River Red Gum is equidistant from the flatlands on which Adelaide is now built and the Murray River—both about 35 kilometres away. It sits in a hidden glen, 10 metres above a bubbling creek, and looks almost due east towards the Murray River. On a clear day the green trees following the course of the great river are easy to see. The tree is surrounded by she-oaks, other small eucalypts and wattle in bloom.

To visit this tree on a windswept winter's evening, with a chill in the air and clouds scudding overhead, is a chance to immerse oneself in Aboriginal history and imagine the magic of the Dreamtime. These photos were taken in the early morning, the rising sun bathing the tree in a soft glow.

SUGAR GUM

Eucalyptus cladocalyx

Waite Arboretum
Adelaide, South Australia

AMERICAN PRESIDENT JOHN F
Kennedy was fond of telling
the story of the great French
soldier and administrator
Marshall Lyautey, who once
instructed his gardener to plant
a tree. When his gardener responded that the tree would
not reach maturity for 100 years, Lyautey urged, 'In that
case, there is no time to lose; plant it this afternoon!'

Such thoughts must surely have crossed the minds
of those who, over the years, have considered that most
far-sighted of projects—an arboretum. We should be
thankful that all have, by definition, shared Lyautey's
approach rather than his gardener's. Just because they
would never see the full fruits of their labours was no
reason not to proceed.

South Australian businessman, pastoralist and bene-
factor Peter Waite was such a person. In 1914 he donated
to the University of Adelaide *Urrbrae* estate, a house and
gardens set in 54 hectares in the foothills of Adelaide,
together with funds to establish a research institute.
He specified that 27 hectares be set aside as a park or
garden in perpetuity for the enjoyment of the public.

The property passed to the university in 1922 on the
deaths of Waite and his wife Matilda, and planting of
the arboretum began in 1928 under the watchful eye
of Professor Arnold Richardson, the first director of
the Waite Agricultural Research Institute. The second
director, Professor James Prescott—a soil scientist—
was particularly interested in exotic species from
countries with climates similar to South Australia's.

Today the Waite Arboretum contains 2200 trees, rep-
resenting more than 800 species. Each tree is labelled and
mapped. Special collections include eucalypts, of which
there are 360 different kinds, pears, banksias and oaks.

There are also several Dragon Trees from the Canary
Islands, Morocco, Madeira, Cape Verde and the Azores,
Torrey Pine from California, Cyprian Cedar from
Cyprus, *Cupressus dupreziana* from Central Sahara
and *Pyrus syriaca* from Asia Minor.

Arguably Waite's greatest arboreal legacy is not the
trees in the arboretum, but a dozen soaring Sugar Gums
that line the first third of the driveway which snakes its
way from the main road to Waite's old residence. The
gums, planted in 1877, serve as a natural avenue, through
which visitors must pass in order to reach both *Urrbrae
House* and the arboretum.

None is more stately than the easternmost Sugar Gum,
which sits apart from the others on the southern side
of the driveway, set against the backdrop of the Adelaide
Hills. The tree, 6 metres in circumference and 40 metres
high, watches over the arboretum's conifer section, which
includes fine examples of Bunya Pine, Cedar of Lebanon,
Hoop Pine, Himalayan Cedar, Golden Deodar Cedar
and Canary Island Pine.

Sugar Gums are endemic to South Australia, where
they occur in three disjunct areas: the southern Flinders
Ranges, Kangaroo Island and the Eyre Peninsula.
They have been particularly popular as windbreaks
and shelters, and can be seen surrounding wheat fields
throughout Victoria, South Australia and Western
Australia. Their trunks are smooth and mottled with
patches of off-white, yellow and bluish-grey.

It is extraordinarily relaxing to sit under this
magnificent Sugar Gum on a spring morning, with
students wandering past on foot or bicycle and a
cacophony of bird noises overhead. There are squawks
and shrieks from Rosellas, Galahs, Long-billed Corellas,
Rainbow Lorikeets and Sulphur-crested Cockatoos.
Old Sugar Gums are known for their hollows and
holes, so it is likely this tree is also home to
Kookaburras and Tawny Frogmouths.

PRIVATE TREES

MONKEY PUZZLE

Araucaria araucana

Alton
Mount Macedon, Victoria

BETWEEN THE 1840s AND THE 1860s THE CONSTANT SCREAM OF sawmills broke through the still air of Mount Macedon, 60 kilometres north of Melbourne. The mill-owners were harvesting the mountain's giant eucalypts and selling the timber for fences, houses, railway sleepers and as planks to shore up mine shafts.

By the time the gold rush ended, Mount Macedon was almost bare. The Victorian government, fearing for the mountain's future, offered 10-acre lots for sale, provided the owners planted trees on their plots.

The government's concern was understandable because Mount Macedon had a lot going for it. It was close to the burgeoning city of Melbourne—which had grown significantly in size and wealth during the gold rush—and its cold climate was popular with Northern Hemisphere immigrants looking to escape the harsh summer heat.

In 1874 banker Sir George Verdon returned from England, where he had been Victoria's Agent-General, and bought a 10-acre block just below the mountain's 1011-metre summit. He set about building a two-storey replica of an English hunting lodge, clad in red terracotta shingles, which he named *Alton* after *Alton Towers*, his ancestral home in England.

Among the Blackwoods and Stringybarks he planted many exotic trees, under the guidance of Melbourne Botanic Gardens directors William Guilfoyle and Ferdinand von Mueller. Twenty years after the garden was designed, the *Gisborne Gazette* said it possessed 'oaks, ashes, elms, sycamores, poplars and various conifers rearing their stately heads among the native trees'.

Take the zigzag path below *Alton*'s tennis court and you will see, looming above garden beds of azaleas and rhododendrons, a 21-metre Monkey Puzzle, most likely planted around 1890 from an imported seedling.

Life has not necessarily been easy for this grand tree. Winters at Mount Macedon can be very cold, and the 1983 Ash Wednesday bushfire careered up the mountain from the south, burning much of *Alton*'s garden.

Although Monkey Puzzles are rare, and can be difficult to grow in Australia, von Mueller distributed many *Araucarias*. He was particularly fond of mainland Australia's two *Araucarias*—the Bunya Pine and the Hoop Pine. Bunya Pines in particular are often confused with Monkey Puzzles.

Although there are eight of them at *Alton*, head gardener Adrian Woollard describes this Monkey Puzzle as the 'outstanding specimen', largely because of its canopy. 'Most Monkey Puzzles in their natural habitat grow branches up high and have bare trunks down low, but this has branches growing to the ground, which is quite rare,' he says.

Monkey Puzzles are evergreens, like all *Araucarias*, and are the hardiest *Araucaria* species. The leaves are thick and tough, up to 5 centimetres long, with sharp edges. They are also dioecious—the species has separate male and female plants—and rely on the wind for pollination.

Monkey Puzzles are native to Chile and west-central Argentina. In recent decades the natural forests in South

THE FIRST CONFIRMED INTRODUCTION OF THE MONKEY PUZZLE TO EUROPE WAS IN 1795 THROUGH ARCHIBALD MENZIES, AN EARLY EXPLORER OF THE WEST COAST OF THE AMERICAS. MENZIES HAD BEEN ENTERTAINED BY THE GOVERNOR OF CHILE IN 1793 AND HAD POCKETED A FEW OF THE NUTS SERVED AT DINNER. HE LATER SOWED THEM ON BOARD HIS SHIP.

America have been dramatically reduced and the species is now considered vulnerable. The seeds are an important food source for the Indigenous people and among some tribes they are considered to be an aphrodisiac when finely ground.

The first confirmed introduction to Europe was in 1795 through Archibald Menzies, an early explorer of the west coast of the Americas. Menzies had been entertained by the Governor of Chile in 1793 and had pocketed a few of the nuts served at dinner. He later sowed them on board his ship. Five plants survived and two of them were given to Kew Gardens (the last of them dying in 1892). In 1972 a reputed original was still growing at Holker Hall in the Lake District.

In Victorian times the trees became fashionable throughout Europe. It is thought the name came from Britain around 1850, when the species was not well known and had no popular name. Legend has it an owner of a tree in Cornwall showed it to some friends and one remarked, 'It would puzzle a monkey to climb that.' The name Monkey Puzzler was adopted, which became Monkey Puzzle.

After Verdon's death in 1896 *Alton* was bought by Robert Whiting, the owner of the neighbouring property *Hascombe*. In 1926 Whiting sold *Alton* to George Nicholas—pharmaceutical chemist, amateur geologist and botanist. During the Nicholas' family ownership of *Alton*, the gardens surrounding the house were extended and many interesting and rare trees were planted.

Thankfully, *Alton* survived the 1983 bushfires that destroyed almost 400 homes on Mount Macedon. As the fire swept towards the house, *Alton*'s manager Steve Swan doused the house from taps in the garden and the horse trough. The wind changed and the fire stopped 3 metres from the house.

Today *Alton* is owned by the Eshuys family. Its garden remains one of Australia's finest, and includes more than 600 trees and several thousand plants. Twenty-four of *Alton*'s trees are listed in the *National Trust Register of Significant Trees*, including Coast Redwoods, a Red Maple, Spanish and Douglas Firs, Sweet Chestnut and a Variegated Sweet Chestnut, an Oro Oro (*Nestegis montana*, an obscure tree from New Zealand), a Western Red Cedar and several species of spruce.

LONDON PLANE
Platanus × acerifolia

Melbourne Club
Melbourne, Victoria

ON 23 MARCH 1995 A GROUP OF MELBOURNE CLUB MEMBERS met for dinner in Melbourne's Collins Street to feast on 'grilled darne of Atlantic salmon, pan fried supreme of chicken served with pesto and a light chocolate bavarois served with a chocolate plane tree leaf and garnished with fresh berries'.

They had gathered to toast the health, and celebrate the hundredth anniversary, of the club's great London Plane, which dominates the lawn in the garden.

For as long as any could remember, the tree had been a source of shade from the summer sun, and its long, gravity-defying limbs a source of great wonder. The tree is as much a part of Melbourne Club life as port and cigars. 'Let's meet under the Plane' is an oft-quoted arrangement.

At the dinner Professor Carrick Chambers, director of Sydney's Royal Botanic Gardens, sang the tree's praises and the group heard 'Ode to the Plane Tree' from Handel's famous Largo, sung by Xerxes, King of Persia. Xerxes, much to the consternation of those around him, had fallen in love with a Plane while marching to battle in 470 BC.

May fate, ye tender and ye beauteous leaves
Of my beloved plane, to you prove kind.
May thy dear peace be undisturb'd by storms,
By thunder's rage or by the lightning's blast:
Nor mayst thou be,
By ruffling winds, profaned.

Legend has it that Xerxes was on the way to invade Greece with a vast army when he came upon the Plane. He became so enchanted that he stopped his army and presented the tree with golden ornaments and assigned it a personal bodyguard. He had a gold medal engraved with the image of the tree, which he wore as an amulet.

In 1844 John Claudius Loudon, the great horticultural writer, wrote that Xerxes' infatuation lasted for several days, and he became 'entirely oblivious of his army'. The Greeks later defeated the Persian king's army, the loss attributed to the delay and to Xerxes' love-scrambled mind.

While most trees are planted to beautify a garden, or provide a habitat for birds, or shade, this tree was planted

LEGEND HAS IT THAT XERXES WAS ON THE WAY TO INVADE GREECE WITH A VAST ARMY WHEN HE CAME UPON THE PLANE. HE BECAME SO ENCHANTED THAT HE STOPPED HIS ARMY AND PRESENTED THE TREE WITH GOLDEN ORNAMENTS AND ASSIGNED IT A PERSONAL BODYGUARD.

for another reason. It is believed that a Melbourne Club member, or members, planted the tree following a fare-well ball in 1895 for the Victorian Governor, the Earl of Hopetoun, and his wife, the Countess of Hopetoun. The aim of the members was to prevent a marquee from ever being pitched again in the club's garden.

One member, JC Mackinnon, later stated that the tree was planted before the ball and the marquee was pitched over it. Photographs from 1905 show the tree at a height of 3 metres.

Of the three most common planes, the Melbourne Club's tree is the least common. The Oriental Plane (*Platanas orientalis*) and the American Sycamore (*Platanus occidentalis*) are the best-known species, and have been cultivated widely. The London Plane (*Platanus × acerifolia*) is a hybrid of these two, grown from seed at Oxford Botanic Gardens in the 1780s.

The plane has a long history as a major tree in cultivation, from early groves in Athens to the lines of trees that dominate cities such as London, New York, Philadelphia, Melbourne and Sydney. It has a high tolerance of pollution, drought and disease; it sheds its outer layer of bark, ridding itself of soot and other pollutants that can harbour pests in other species. It has also been cultivated in far-flung places like Iran, Kashmir and Crete. Its leaf resembles that of a sycamore or maple.

The pressure on the Melbourne Club's tree has, over the years, caused it some stress. Concerned members have met, discussed its health and then lathered it with feed, aerated its soil and judicially pruned its limbs. History has it that a group met once to discuss the tree's state and from its fringe a farmer from the Riverina was heard to mutter under his breath, 'Ever thought of water?'

COOLIBAH
Eucalyptus coolabah

Monkira Station
South-west Queensland

BROODING ON AN OLD WATERHOLE IN QUEENSLAND'S famous Channel Country, on the floodplains of the mighty Diamantina River, is Australia's largest Coolibah. The locals call it the Monkira Monster, a name to strike fear into the heart of any child at story time.

It is certainly one of Australia's less accessible great trees. To get there you must fly to Windorah or Bedourie in western Queensland and then drive for the best part of two hours through flat, lonely country interrupted only by stone monoliths and the odd gate pointing down dusty and seemingly endless desert tracks. The tree is half an hour's drive from the nearest public road.

The journey is worth it.

The tree, with a trunk circumference of more than 10 metres, lives on Nueragully Waterhole on *Monkira*, a one-million-acre cattle station. The waterhole—truth be told—is more a lake, especially when good rains fall to the north and water flows south towards Lake Eyre. Some of the water flows down Millawarrina Creek and fills the waterhole.

In good years the Monkira Monster's trunk can be submerged to a depth of 2 or 3 metres. In other years, like in 2002, the waterhole dries up completely and the tree is left high and dry, like a shipwreck on a shore.

The tree, battered and crumbling, tells of the tribulations of trying to survive in the Channel Country, where the good years are wonderful and the bad years full of challenges.

It's unlikely that Australian poet, jockey and politician Adam Lindsay Gordon ever clapped eyes on the Monkira Monster, but the words he once wrote about an old gum tree are very apt here:

> *When the gnarl'd, knotted trunks eucalyptian,*
> *Seem carved, like weird columns Egyptian...*

The tree is protected from easterly winds by a large sand dune and its position on the east side of the waterhole allows it to catch the full strength of the afternoon sun. This confluence of fate has doubtless assisted it to survive so long and grow to such an impressive size.

For as long as anyone can remember, this tree has been big. The National Library of Australia has a grainy black and white photo taken in 1952 by Arthur Groom—photographer, author and conservationist—of the

manager of *Monkira*, Bob Gunther, sitting on one of the tree's stupendous lower branches. Gunther, enveloped by the tree's bulk, looks out of the photo with a penetrating gaze. Twelve years later author Albert Brooks wrote about the Monkira Monster in his book *Tree Wonders of Australi*a.

But the past 60 years have not been kind. Today the tree's rough bark crumbles to the touch and it has lost many significant branches, causing great rips and wounds in its trunk. Other branches have died but remain attached to the trunk, offering fine support for myriad spider webs.

Many of the Monster's giant branches—some large enough to be trees in their own right—disappear into the silt of the waterhole and re-emerge up to 30 metres away. With all the appearance of a dying Kraken, these semi-escaping limbs sport healthy collections of leaves.

In September 2008, 82-year-old former wheat farmer Lindsay Dunstan made the six-day 4000-kilometre round trip—from Murwillumbah in northern New South Wales to *Monkira*—to pay his respects to the tree after reading about it in a book. 'It was well worth the trip. The tree is very impressive,' he says. 'I paced around the tree's drip line and it was more than 200 metres in circumference, which is amazing. It really is a straggly and grotesque giant, and probably not the most attractive tree under the sun. It's beginning to break apart a bit, but what a wonderful life it's had.'

We pay our own visit to the Monster in April, as the water recedes after good summer rains. We spend an hour marvelling at its size and climbing on its branches. Ducks paddle unhurriedly nearby as the sun sets and the Monkira Monster takes on a warm earthy glow, as it has done many thousands of times before.

ENGLISH OAK

Quercus robur

Connorville
Near Launceston, Tasmania

OF THE THINGS RODERIC O'CONNOR BROUGHT WITH HIM when he sailed from Ireland to Van Diemen's Land in 1824, aboard his own ship the *Ardent*, aiming to start life afresh in the new colony with his sons Arthur and William, it is debatable whether anything had more meaning for him, or more long-term prospects, than a small acorn he carried in his pocket.

Not long afterwards, when Governor George Arthur had granted O'Connor 400 hectares under the Great Western Tiers 50 kilometres south of Launceston, the first thing O'Connor did was dig a hole and plant the acorn. The resulting English Oak is today one of the finest trees in Tasmania.

The acorn was as much a metaphor for O'Connor as it was part of his attempt to re-create an Irish garden in his new home. It said volumes about a man willing to take on the risks of starting afresh in a new and untried colony. The tree, with its strong roots and sturdy limbs, seemed to embody O'Connor's gimlet-eyed approach to a new beginning.

Within four years O'Connor had bought 1300 hectares, which carried 100 cattle and 2500 sheep. By the time he was buried in Tasmania's dark soil in 1860, he had amassed 30 000 hectares, and his descendants still farm the land today.

Today, the English Oak at *Connorville* is as imposing as one of the earliest introduced trees in the state should be. It stands 22 metres high, measures 5 metres around the base, and has a fine canopy 31 metres in diameter.

It receives much love and affection, keenly tended by the sixth generation of the O'Connor clan to run *Connorville*—another Roderic O'Connor—and his wife Kate. Seven years ago they gave birth to a son, Lachlan. If the tree was ever assigned the job of

IT IS DEBATABLE WHETHER
ANYTHING HAD MORE MEANING
FOR HIM, OR MORE LONG-TERM
PROSPECTS, THAN A SMALL ACORN
HE CARRIED IN HIS POCKET.

ensuring the family line continued, and it may well have been, it has been successful. It seems only fair that as long as it continues to bud, an O'Connor will run *Connorville*.

The English Oak is situated to the north of the main house—a 1922 homestead—in a garden that also includes Japanese Maple, Douglas Fir, Copper Beech, ash, willow, poplar, Ponderosa Pine, Monterey Cypress and a Golden Elm planted by the Duke of Edinburgh in 1954.

And it has played a major part in the lives of many O'Connors. Where better to have a cocktail party but under the tree's sweeping boughs? Where else would a magician choose to entertain a group of children for an hour on a summer's evening?

Connorville is, by any measure, an extraordinary place. From the old oak a road leads through a village that includes a saddlery, butchery, baker, blacksmith's shop, a three-storey mill, single-men's quarters, killing shed, stables, coach house, chaff house, library and a store-keeper's building, from where rum was rationed to the workers when it was the accepted currency.

Back in the mid-1800s, the village was almost feudal. The hundred or so employees were mainly ticket-of-leave men (ex-convicts) assigned to the original Roderic O'Connor in the 1830s. Hundreds of people have worked at *Connorville* since, and each has been looked over by a faithful old oak, part of one man's vision and one family's history.

PONDEROSA PINE

Pinus ponderosa

Khancoban Station
South-east New South Wales

IN 1803 US PRESIDENT THOMAS Jefferson paid the French government a shade over US$23 million for 2.15 million square kilometres of what is now central United States—a deal known as the Louisiana Purchase. The area comprised around 23 per cent of today's United States of America.

Although he originally made the purchase because he did not want France and Spain having the power to block American access to the port of New Orleans, Jefferson was aware that there could be other benefits in the unknown land out west, so he persuaded Congress to allocate US$2500 for an expedition.

The following year two men of sturdy disposition, Meriwether Lewis and William Clark, set off for the west coast. Along the way, in addition to crossing the newly bought territory, including the formidable Rocky Mountains, they observed and described 178 plants and 122 animals.

Among the trees they saw was the fantastic Ponderosa Pine, the dominant pines of the Rocky Mountains, which stretch from Canada to Mexico. In fact, they knew the trees were there weeks before they saw them in Montana, having seen pine cones bobbing along the Missouri River in the Dakotas, hundreds of miles downstream.

Ponderosa Pines have been planted widely throughout the world. They were introduced to Australasia for two reasons—for timber and as ornamental trees for botanic gardens.

It is hard to imagine any Ponderosa Pine in Australia occupying a more dominant location than the famous tree at *Khancoban Station* in south-east New South Wales. It sits high on a ridge protruding into the floodplain of the upper Murray River system. In the winter the Snowy Mountains are a symphony of white, and in the springtime the bright yellow of the nearby canola fields is almost blinding. The tree, more than 40 metres tall, is visible for more than 10 kilometres, from well beyond Khancoban township.

How did this tree come to be here? It is a question that *Khancoban Station*'s current owners cannot answer. Arborist Alex Bicknell, who has tended the tree since 2004, estimates its age at 130 years. If it had been planted in 1875—which is possible—that would coincide with the purchase of the property by Thomas Mitchell. Mitchell passed *Khancoban Station* to his son John Francis Huon Mitchell, who later bequeathed it to his sister Emma in 1921.

The pine is a similar age to the seven Coast Redwoods, which also hail from north-west United States, that live in a neat line next to the homestead. Dotted around this part of the property, from the same era, are also elms,

THE OLDEST LIVING PONDEROSA PINE IN COLORADO IS 850 YEARS OLD. ONE, IN COLORADO'S MESA VERDE NATIONAL PARK, LIVED FOR 1047 YEARS.

planes, oaks, a Canary Island Pine, Blue Atlas Cedar, Golden Ash, Spanish and Douglas Firs, a Tulip Tree and a rare Bishop Pine.

In the past five years 662 trees have been planted at *Khancoban Station*, including several species of oak, Hybrid Strawberry Trees, Bottle Trees and more Golden Ash and Tulip Trees. Because of frosty winters, hot summers, high rainfall and the protected location, there are few other microclimates in Australia where such trees grow as fast, or as well.

Years ago there was a cottage near the top of the ridge upon which the Ponderosa Pine now sits. In addition there were also a dozen Bishop Pines and some Kurrajongs. The Ponderosa is the only survivor; its roots may have found an old terracotta irrigation channel that once ran water around the side of the hill.

Also known as Western Yellow Pines, Ponderosa Pines are not common in Australia, but where they prosper the rainfall is generally high, such as in Victoria's Dandenong Ranges and Mount Macedon.

In the southern Rocky Mountains of the United States, these trees grow in sunny and dry locations at altitudes between 1804 and 2560 metres. They can sometimes attain heights of 65 metres. They can endure drought and high temperatures, and can live beyond 250 years. The oldest living Ponderosa Pine in Colorado is 850 years old. One, in Colorado's Mesa Verde National Park, lived for 1047 years.

COAST REDWOOD
Sequoia sempervirens

Jephcott Arboretum
Near Tintaldra, New South Wales

EDWIN JEPHCOTT, A SEEDSMAN FROM COVENTRY IN ENGLAND, was nothing if not far-sighted. In the mid-1860s he planted 300 exotic trees on 6 hectares on the banks of the upper Murray River, between today's towns of Tintaldra and Jingellic.

Included in his plantings were fifty species of pine, ten species of oak, various planes, ash, and a rare Shagbark Hickory from the Everglades in Florida. He also planted three Coast Redwoods—the world's tallest tree species.

Old man Jephcott originally migrated to Australia to take a position at the Brisbane Botanic Gardens, but he was later persuaded by his cousin to come to Victoria. Jephcott loaded his pregnant wife and four children into a bullock dray and made the long journey south.

He built a modest house overlooking the Murray River—the chimney of which is still standing—and started his extraordinary collection. Importantly Jephcott communicated with, and probably swapped seeds with, Melbourne's Botanic Gardens director Ferdinand von Mueller, who visited Jephcott's arboretum at least three times and doubtless passed on some useful advice.

Many spectacular trees remain at the arboretum, but none is more impressive than the tallest Coast Redwood, which sits in glorious isolation about 100 metres from the Murray River. Could it be the world's only Coast Redwood surrounded, at a discrete distance, by River Red Gums?

The tree is 46 metres tall, and nearly 9 metres around the base. It is probably the largest Coast Redwood in Australia. It has thrived in the foggy microclimate of the upper Murray valley, which resembles that of the US north-west coast, where it originates and where it thrives in the cool, damp air.

But all things are relative. The tree is a pipsqueak compared to the monsters in the United States, notably California and Oregon. There the Redwoods must outcompete each other for light, and many reach heights of more than 100 metres.

COULD THIS BE THE WORLD'S ONLY
COAST REDWOOD SURROUNDED, AT
A DISCRETE DISTANCE, BY RIVER
RED GUMS?

The world's tallest living Coast Redwood, and the world's tallest tree, was discovered in 2006 in the Redwood National Park in northern California. It is 115.6 metres tall and estimates are that it holds 500 cubic metres of wood.

Today the Jephcott Arboretum is overseen by the sixth generation of Jephcotts. In addition to being a valuable collection of trees, it may hold some valuable genetic secrets. Three pines in the arboretum, despite much study from botanic experts, have yet to be identified, so may hold vital clues to some lost species.

Edwin Jephcott's son Sydney was also keen on trees, as well as being one of Australia's bush poets. He was asked once to name his favourite tree in the arboretum and responded that the choice was too difficult. 'They must be viewed like mistresses: one for each season,' he said.

THE TREE HAS THRIVED IN THE FOGGY MICROCLIMATE OF THE UPPER MURRAY VALLEY, WHICH RESEMBLES THAT OF THE US NORTH-WEST COAST.

LEMON-SCENTED GUM
Corymbia citriodora

Nallama
Near Tocumwal, New South Wales

WHEN 74-YEAR-OLD HARRIET HUNT, DAUGHTER OF WILLIAM and Maria Hunt of Bristol, England, died at *Nallama*—15 kilometres west of Tocumwal in southern New South Wales—on 30 September 1879, a Lemon-scented Gum sapling was trying desperately to eke out a future in the sandy soil of the homestead's garden.

Today that tree, around 130 years old, is one of the finest examples in the country of this much loved native. Its trunk is 4 metres in circumference and its knotted limbs reach a height of 25 metres. Its canopy spans 35 metres. It has good company; nearby is a purple Jacaranda, an Olive and a gnarly pine. A windmill cranks away near the homestead, providing bore water fit enough to drink.

Greg Ingram, who manages *Nallama*, says to stand near the tree when the sun comes out after a good storm is to experience sensory overload. The lemon smell is intoxicating. This also happens when he mows the grass under the tree. The tree's seeds and leaves are crushed by the mower's blades, releasing a citrus smell strong enough to wrinkle the nose.

The past three years have not been kind to the tree. Rainfall at *Nallama*—a 600-hectare sheep and wheat property—is normally 400 millimetres a year, but the last three years have each yielded little more than 200 millimetres. In good years the boughs of the Lemon-scented Gum bear so many leaves that they touch the ground. Today the leaves are a little sparse. We visit it after 50 millimetres of rain, and the tree looks content, its new green leaves, interspersed with flashings of red, contrasting against a deep blue Riverina sky.

Each summer the tree sheds its rough, pink bark. The new bark on the tree's trunk and major branches is a bright, smooth yellow. It is a brilliant—and there is no better word for it—example of a tree's appearance changing with the seasons.

Lemon-scented Gums are graceful trees, with long flowing branches and fine canopies. One of their great strengths is their adaptability. They travel well from their native region between Mackay and Maryborough in Queensland. They are not only popular because of their appearance and smell, but the oil in their leaves is rich in citronella, so they have a commercial use as well.

Today Harriet Hunt lies next to her brother, John Hawkins Hunt (who died aged eighty-six), in a sandy grave 500 metres south of the homestead. Not surprisingly their shared grave, surrounded by three old gum trees, sits on a small rise and faces the homestead, giving them a wonderful view of the glorious Lemon-scented Gum. When the hot wind blows from the north, as it often does around here, they might even get a whiff of lemons.

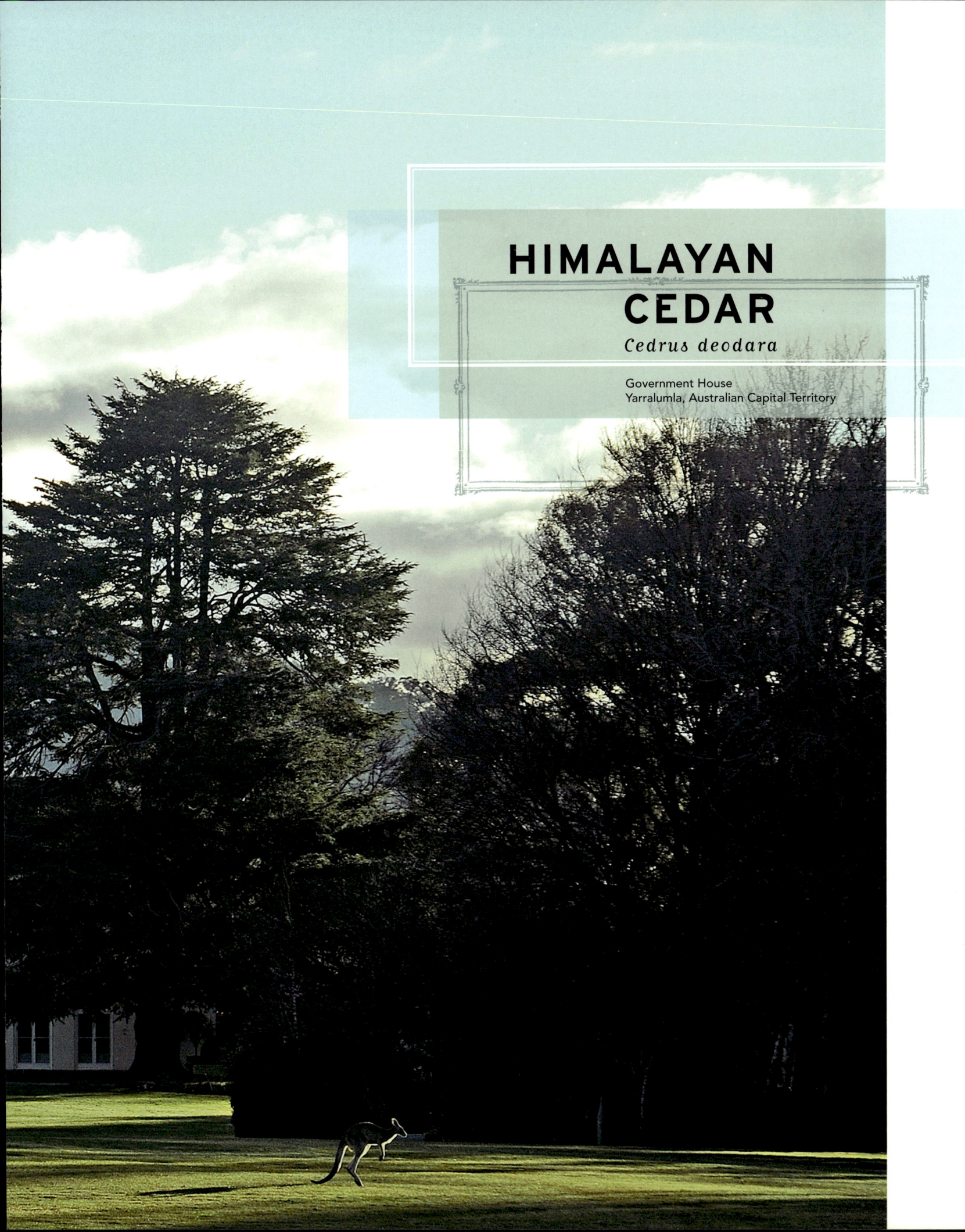

HIMALAYAN CEDAR

Cedrus deodara

Government House
Yarralumla, Australian Capital Territory

MANY STRANGE THINGS were shipped to Australia from Britain during the mid-1800s, but few—from a botanical point of view at least—were more unusual than two 5-year-old Himalayan Cedars, which were offloaded on a Sydney wharf in 1837.

One sapling made a short journey east and was planted in the garden at *Vaucluse House.* The other was transported to *Yarralumla,* a sprawling sheep station in New South Wales' south-east. The station's owner, Terrence Murray, was about to get married and had ordered his gardener, Terrence McHugh, to create a European garden for his bride-to-be, Mary Gibbes.

Water was not a problem—a river ran close to the house—so McHugh did not hold back, ordering and planting hundreds of trees, both native and exotic. And, he reasoned, no self-respecting European-style garden in Australia was without a decent Himalayan Cedar.

Over the years many important people have planted trees at *Yarralumla,* today the home of Australia's Governor-General in Canberra. Those who have wielded the custom-made silver spade include Queen Elizabeth II (Silvertop Gum, 1954), Princess Diana (Golden Ash, 1985), Queen Beatrix of the Netherlands (Mugga Ironbark, 1988) and Bill Clinton (White Dogwood, 1996).

But none of these trees is likely ever to match the grandeur of McHugh's Himalayan Cedar, now more than 170 years old. The tree occupies a commanding position next to Government House and can be seen from virtually anywhere on the 54-hectare property. It dwarfs the house, which is by no means small, and it dominates the view up the driveway. It has, therefore, witnessed the arrival and departure of countless foreign dignitaries, as well as plenty of Australian history: the calling of elections, the making of policy and the hatching of plots to oust prime ministers.

Like most old and large trees, it has dropped some branches over time, including one that almost crushed Bill Hayden's son's car. Today the tree—8 metres in circumference at the base, 30 metres high with a 25-metre-wide canopy—is in very good health, and is probably the most impressive Himalayan Cedar in Australia.

Mind you, it should be healthy, with nine full-time Government House gardeners tending to its every need. It had a major pruning in 2006, when seven tonnes of

THE SEEDS HAVE BEEN USED TO
GROW MOST OF THE HIMALAYAN
CEDARS IN THE CANBERRA DISTRICT,
INCLUDING 68 000 TREES ON
NEARBY MOUNT STROMLO AND
GREEN HILLS, PLANTED BETWEEN
1915 AND 1920.

foliage and dead wood were removed, mainly from the crown and larger branches. One of its major limbs is held up by a steel cable, but that is the only sign of advancing age. There is no reason it will not live for another 100 years. It is already the oldest exotic tree in the garden, which boasts 4500 trees and more than 180 species.

Himalayan Cedars, also known as Deodars, are native to the western Himalaya, and can be found in Afghanistan, Pakistan, Indian, southern Tibet and western Nepal.

The species has adapted well in western Europe and around the Black Sea, as well as through southern and central China and on the west coast and in the south-east of the United States. Its Himalayan origins explain why this tree has survived so well in Australia's capital, which has cold winters and even rainfall.

Cockatoos are this tree's biggest enemy, regularly feasting on its seeds. Every two years, for three months at a time, *Yarralumla*'s gardeners put cages around the tree's cones and harvest the seeds. The seeds have been used to grow most of the Himalayan Cedars in the Canberra district, including 68 000 trees on nearby Mount Stromlo and Green Hills, planted between 1915 and 1920.

In the 1940s the Duke of Gloucester, the Governor-General at the time, called upon Lindsay Pryor, director of Canberra's Parks and Gardens (which looked after *Yarralumla*'s gardens), to open up the vista from the main house to the east. Pryor 'feared a massacre' and was worried especially about the future of the great tree. He steered activity to some old Radiata Pines in the area.

The Himalayan Cedar survived. So, incidentally, did Pryor.

HYBRID STRAWBERRY TREE

Arbutus × andrachnoides

Raeburn
Near Goulburn, New South Wales

LIKE SOME HORSES and dogs, half-breeds of trees often perform better than their pure-breed parents. The Hybrid Strawberry Tree is part Cyprus Strawberry Tree (*Arbutus andrachne*) and part Irish Strawberry Tree (*Arbutus unedo*). The hybrid grows faster, and is hardier, than both parents. It occurs naturally in the wild and is common in Greece, where the two species meet.

The tree, an evergreen, also travels well. Twenty kilometres west of Goulburn in New South Wales, near the small hamlet of Breadalbane, is an exceptional Hybrid Strawberry Tree—one of the grandest in the country. It sits in the middle of the circular driveway of the stately home of *Raeburn*, now owned by the Taylor Family. The region was settled in 1837, and the tree would have been planted in the decade or two that followed, probably from a seed from Melbourne, Sydney or Adelaide botanic gardens.

This used to be exceptional sheep country, but today it is in the grip of a drought. The surrounding countryside is dry as a crisp, but the tree is in good health.

We visit it in mid-winter, in a snowstorm. The tree's branches writhe against the forbidding sky, the darkness only serving to bring out its exceptional colours. The bark of the Hybrid Strawberry Tree is more colourful than almost any other tree's, and changes with the seasons—through yellow, orange, red, pink and brown.

Hybrid Strawberry Trees are relatively rare in Australia, although they had a brief burst of popularity in the mid-1800s, when they were planted as an ornamental tree. The original owner of *Raeburn* owned another five sheep stations, and planted a Hybrid Strawberry Tree at each one.

Those of them that still survive could do so for another century or so. Hybrid Strawberry Trees live for up to 300 years. Generally they do well in Australia, and do not mind the heat and dryness. The climate of the area around Canberra is not dissimilar to that of the mountains of Cyprus. Hybrid Strawberry Trees also thrive in the highlands and tablelands down Australia's east coast, from Toowoomba to Hobart.

The tree is a member of the Ericaceae family which includes thousands of species. Included in the family is heather, which turns the Scottish moors purple each year, and rhododendrons and azaleas. (Australia had only nine species, including its only two rhododendrons, but in the last couple of years plants from other families have been moved to Ericaceae, and now Australia is better represented. A notable new member of it is the Pink Heath (*Epacris impressa*), Victoria's floral emblem.)

The Hybrid Strawberry Tree's flowers are like little bells, not unlike heather, and they emerge in winter. The tree can fruit, but don't expect ever to see it. The fruit takes two years to ripen and birds generally eat it as soon as it appears.

THE BARK OF THE HYBRID STRAWBERRY TREE IS MORE COLOURFUL THAN ALMOST ANY OTHER TREE'S, AND CHANGES WITH THE SEASONS—THROUGH YELLOW, ORANGE, RED, PINK AND BROWN.

LOCAL GIANTS

KARRI
Eucalyptus diversicolor

Warren National Park
Western Australia

AT THE RISK OF STARTING A
national argument, this Karri—
on the edge of the Warren
National Park, 12 kilometres
south-west of Pemberton in
Western Australia—may
well be Australia's most beautiful tree.

The fact that it is 80 metres high—getting towards
the upper limit for a Karri—is impressive enough, but its
beauty lies more in its shape. It has a 6-metre girth and
its trunk is billiard-cue straight for the first 60 metres,
before branching out into a full and balanced canopy.
It is unusual that such a tall tree should be unscathed
by either wind or lightning.

To lie in a forest of Karri in the late afternoon autumn
sunlight and look up at their dizzying heights—the
clouds skimming overhead—is to risk vertigo. It is hard
to take in the perfection of such tall straight trunks,
such giddily high upper branches.

The Karri is the tallest species in Western Australia, but
is restricted to a relatively small high-rainfall area of the
state's south-west, from Albany (400 kilometres south of
Perth) to Cape Leeuwin in the famous Margaret River
wine-growing region. It is a pity that its isolation means
the vast majority of Australians have not walked in a
Karri forest, and had the chance to look up at these giants.

Generally, the branches of mature Karris appear only
in the top third of the tree, so their trunks are long and
spectacular. When the Karri sheds its old bark, its new
bark varies in colour from orange to yellow, grey and
white. However, its name *diversicolor* comes from the
different shades of its leaves—dark green on one side,
light green on the other. The tree has dark green clusters
of leaves hanging from upper branches while the lower
branches put out lighter green clusters. The white flowers

produce excellent honey that is harvested by birds such
as Lorikeets, which then fly off to pollinate other forest
trees. Karris produce small barrel-shaped fruit. For such
huge trees it is interesting to note that they don't live
much beyond 300 years.

The timber of the Karri and the Jarrah (*Eucalyptus
marginata*), which it closely resembles (and which is
also a native of south-west Western Australia), are
among Australia's most valued hardwoods. Both are a
deep red, durable, hard and easily fashioned. And if you
can't tell the difference between Karri and Jarrah, burn
a splinter of wood. The Karri leaves white ash, the
Jarrah grey or black ash.

BECAUSE KARRIS HAVE AN ISOLATED LOCATION—IN WESTERN AUSTRALIA'S FAR SOUTH-WEST—THE VAST MAJORITY OF AUSTRALIANS HAVE NOT WALKED IN A KARRI FOREST. TO LOOK UP AT THE DIZZYING HEIGHT OF A KARRI IS TO RISK VERTIGO.

MOUNTAIN ASH
Eucalyptus regnans

Yarra State Forest
Victoria

AROUND THIRTY YEARS AGO AMATEUR PROSPECTOR WERNER Marschalek was fossicking for gold and gemstones in the thick undergrowth of a Myrtle Beech and Sassafras rainforest south-east of Warburton in Victoria, when he chanced upon a tree that made him stop and stare. The tree's trunk was so large it blotted out the surrounding bush, and its upper branches disappeared into the heavens.

Marschalek stood, gazing up at the Mountain Ash, feeling both awe and apprehension. He had spent enough time in the area to know that the tree was probably the largest in the forest and, consequently, its future was in jeopardy.

Logging was running unchecked in the region, and he reasoned it was only a matter of time before someone from a logging company chanced upon the tree, as he had, and chopped it down. He began to lobby local politicians to have the tree recognised and protected.

Today 'the Ada Tree', as the locals call it (it sits at the headwaters of the Ada River between Powelltown and Noojee), is safe from loggers. A team of people, including, for a while, prisoners from the local jail, have built a path to the tree from the roadhead. A wooden platform protects the tree's roots from compaction. This Mountain Ash has become a great attraction, receiving visitors who come to gaze at its prodigious size.

The tree, probably more than 300 years old, is impressive by any measure. It is 76 metres tall and has a girth of 15.7 metres. Estimates are that it weighs as much as a 747 aeroplane. Its root system probably extends over half a hectare. The Department of Sustainability and Environment says that, before it lost its upper branches, it may well have been 120 metres tall, which would have put it among the world's tallest trees.

When settlers first arrived in Australia's southern states, they discovered giant stands of Mountain Ash in the mountains to Melbourne's east, on Mount Macedon, in the Otway Ranges and in Tasmania. Victoria's were the tallest of these, but decades of land clearing, logging and bushfires—such as those in 1938—have handed that mantle to Tasmania. But perhaps only temporarily. Tasmania's tallest stands of Mountain Ash are nearing the end of their lives, while Victoria has many younger trees that are nearly as tall. Unlike many eucalypts, Mountain Ash have no lignotuber (a swollen woody mass at the base of the stem containing dormant buds) which can produce new shoots after fire or axing. Their only means of survival after fire is by massive seed dispersal to germinate young seedlings in the ash.

Today 'the Ada Tree' sits in a 600-hectare reserve, but the devastation of clear-fell logging is all too apparent nearby. Logging has taken place no more than 1 kilometre away, but today the old logging tracks are gradually growing over and the tree is safe.

THE TREE, PROBABLY MORE THAN 300 YEARS OLD, IS IMPRESSIVE BY ANY MEASURE. IT IS 76 METRES TALL AND HAS A GIRTH OF 15.7 METRES. ESTIMATES ARE THAT IT WEIGHS AS MUCH AS A 747 AEROPLANE.

SHINING GUM
Eucalyptus denticulata

Errinundra Plateau
Victoria

IT IS EXTRAORDINARY TO THINK THAT THIS 700-YEAR-OLD Shining Gum—which lives at an altitude of 1100 metres on the southern side of the Errinundra Plateau in eastern Victoria—began its long life during Europe's Middle Ages. It was already growing by the time the Roman Empire suffered its death throes in the mid-1400s and was firmly established when Christopher Columbus set sail for America. It has lived through the Renaissance, the Reformation and the Industrial Revolution.

Although far removed from those events, it has lived through exciting, changing times in its own location. Despite being well inland, 80 kilometres north of Orbost, it has fine views of Bass Strait and would have witnessed the boats of George Bass and Matthew Flinders gliding past—the first Europeans to chart those treacherous waters in the 1790s.

Although there are other ancient Shining Gums in the region, this tree—which lives 200 metres down Rooty Break Track, off Goonmirk Rocks Road near the sawmill town of Bendoc—is surely the elder statesman. It towers above giant Tree Ferns, Silver Wattle, Mountain Pepper Trees, Sassafras, Blackwood and Black Olive Berries.

It is a crusty veteran, cracked with age lines of wisdom and bumps of knowledge. Almost the entire bottom half of the massive trunk, 15 metres in circumference, is a riot of burls. These have been formed over the years as a protection against grubs, which have bored into the tree to find sap. One burl is 3 metres in diameter.

Until 20 years ago it was thought that these gums on the Errinundra Plateau were *Eucalyptus nitens*, the same as the Shining Gums that grow in the mountains of eastern Victoria and New South Wales.

However, in the early 1990s, Professor Pauline Ladiges, then deputy head of the Melbourne University School of Botany, and Masters student Ian Cook decided to explore this assumption. After much research they determined that the Errinundra Plateau trees, as well as others near Mount Baw Baw, differed from 'traditional' Shining Gums in several significant ways, even at a genetic level. Foresters had already noted in their field trials that there were differences in rates of growth and frost tolerance. More bizarrely, the leaves on the Errinundra trees had tooth-like edges, unlike the smooth-edged leaves of *Eucalyptus nitens* and almost all other eucalypts.

The two academics published their findings in 1991 in the journal *Australian Systematic Botany*, and described a new species of Shining Gum—*Eucalyptus denticulata* (dentate: having a tooth-like margin).

The two species are very similar, but why would one species develop tooth-like edges to its leaves? Dr Kevin Thiele, the head of the Western Australian Herbarium and who lived near the Errinundra Plateau for many years, has a theory. He says the leaves of trees growing in damp areas are susceptible to fungi, which thrive in the moisture that collects on the leaves via rain and mist.

IT IS A CRUSTY VETERAN, CRACKED WITH AGE LINES OF WISDOM AND BUMPS OF KNOWLEDGE.

Leaves with smooth edges tend to stay moist for longer and so suffer more from infection. However, water drips more readily from a leaf with a jagged edge, leaving the leaf drier and less vulnerable to fungal attack.

Make the pilgrimage to the Errinundra Plateau and this theory makes great sense. The region is covered with snow for three months of the year and the swirling mist that surrounds the tree for the rest of the year chills the bones.

Why are these eucalypts called Shining Gums, when they appear dusty, unkempt and most *un*shiny? In fact the term (in Latin *nitens* means shining, polished or bright) applies to the trees' leaves, buds and fruit, not just their bark. The fruit of the Shining Gum in particular is distinctively glossy, and looks almost varnished.

This venerable gum tree is not nearly as tall today as it once was. Long ago it lost most of its top, probably to lightning. The Errinundra Plateau attracts much lightning because of its altitude and the fact that the

plateau is largely granite. Before then the tree was probably close to 70 metres tall.

The conditions that have allowed the tree, and others around it, to reach such a grand age—cool weather, the south-facing location and the numerous damp gullies nearby—may also lead to the eventual disappearance of the species in this area. Many eucalypts need large fires to regenerate but as long as the damp conditions continue, such fires are unlikely. These ancient Shining Gums will eventually die, probably within the next 100 years, without ever having had the chance to propagate their seeds. The rainforest that currently lives beneath the canopy will in time take over.

Undoubtedly this tree has had a charmed life, largely because of its size. By the time loggers arrived in these parts it was already too big for their bullock drays, so they left it alone. The region is now part of the Errinundra National Park, so this Shining Gum and the trees around it are protected.

BULL KAURI
Agathis microstachya

Lake Barrine
Atherton Tablelands, Queensland

THE STATUESQUE KAURI PINE USED TO BE FAIRLY COMMON IN Queensland, but it became a victim of its own popularity. Since the early nineteenth century woodworkers have sought its attractive and easy-to-work timber—especially for furniture and tongue-and-groove panelling—and the tree was logged extensively. Luckily rainforest trees are now protected throughout the state.

Today, although Kauris are still seen occasionally in the rainforests, and one species in particular, Queensland Kauri (*Agathis robusta*), has been successfully grown in parks and gardens as far south as Melbourne and Adelaide, they are rarely seen outside Queensland. They are not easy to grow and even if they germinate, their seedlings are susceptible to frost.

On the north-west side of Lake Barrine, a volcanic lake 20 kilometres east of Atherton in the Atherton Tablelands, stand two 50-metre-high specimens of Bull Kauri. These are close relatives of their more famous cousins—the New Zealand Kauri, from where the name derives, and the Queensland Kauri.

The largest of the two Lake Barrine Kauris has a 6-metre girth and its bark shows a distinctive spiral pattern. It is estimated to be 1000 years old. Because the Kauri's bark flakes off, the trees are generally free of vines, and neither lichen nor moss can gain a long-term foothold. Despite this, two giant basket ferns have taken root in the forks of the lower branches.

Kauri Pine, the largest of Australia's thirty-eight conifer species, is one of the oldest tree species on Earth. During the Mesozoic era (250–65 million years ago), relations of today's Kauris looked down upon the dinosaurs. Alongside cycads—giant and ancient palm-like plants—these trees dominated the warm, moist forests.

Today's Kauris, and other conifers, are important in helping us to understand the evolution of land plants. Notably, although the Kauri is a conifer, it has broad leaves rather than needles, enabling young trees to capture as much light as possible in the dark rainforest, improving their chances of survival. Kauris produce both male and female cones; the males disperse pollen on the wind and the larger female cones ripen in December and shed their winged seeds on the wind. (The word *Agathis* comes from the Greek; it means ball of twine and describes nicely the female cones.)

The Bull Kauri is one of only three recorded *Agathis* species in Australia—the others being the Queensland Kauri (*Agathis robusta*) and the Black Kauri (*Agathis atropurpurea*). Bull Kauris are restricted to a small region on the Atherton Tablelands, and typically grow at altitudes of between 400 and 900 metres in areas of high rainfall.

TASMANIAN
MOUNTAIN GUM

Eucalyptus dalrympleana ssp. 'Tasmania'

East of Launceston
North-east Tasmania

south-east of Tasmania's Mount Barrow, down winding Boags Country Road high in the hills, is an 86-metre Tasmanian Mountain Gum that nine years ago was on death row. But when the time came to take down this giant tree, the logger drew breath and spared it the chainsaw.

Nick Viney, who has worked in the timber industry for twenty-five years, was driving an excavator that day on the logging coupe. He says he was glad the tree got a reprieve. 'Mountain Gums very rarely grow that big,' he says. 'They usually grow to about 50 metres in good conditions. This is a one-in-a-million tree, a freak tree.'

Viney was so taken with the tree that he has returned to the area several times since, to show it to people and to photograph it. 'It's an emotional thing for me. I prefer to see trees like this left where they are,' he says.

Nearly a decade later, the tree seems content enough, surrounded by 6-metre-high blue gums, the plantation timber of choice in these parts. Behind it and to the side, like acolytes to a reigning monarch, are a dozen or so other white gums. None are as tall or as impressive. 'You cannot leave a big tree in the middle of nowhere because it will blow over,' Viney says. 'It needs other trees around it. The tree will be fine where it is.'

The tree, 14.5 metres in circumference, sits proud and majestic, ramrod straight with 40 metres of smooth speckled white trunk before the first of its five major branches. It is as if nine years ago the tree straightened its back when it heard the chainsaws and dared them to do it. It is remarkably complete and beautifully shaped.

Tasmanian Mountain Gums are often mistaken for Manna Gums (*Eucalyptus viminalis*), although the latter rarely grow above an altitude of 300 metres. This particular tree sits at an altitude of close to 400 metres. The air in the region is cool and clear and the forest around is thick and damp. The tree's startlingly white trunk contrasts vividly with the dark green forest behind.

The other direction, however, shows nothing but devastation, with twisted and charcoaled trunks and trash scattered about, Hiroshima-like. Most of the wood from these parts goes to one of three local sawmills, then on to Japan as woodchips.

'... THIS IS A ONE-IN-A-MILLION TREE, A FREAK TREE.' — NICK VINEY

MOUNTAIN ASH

Eucalyptus regnans

Near Geeveston
South-west of Hobart, Tasmania

WHEN IT COMES TO NATURE, SIZE matters, which is probably why this Mountain Ash—on the Arve Loop Road, near Geeveston, ninety minutes' drive south-west of Hobart—is still with us.

It has lived for close to half a millennium, preceding European settlement in Tasmania by more than 250 years. It has made the most of its damp location and flourished. It lives on the side of a hill and not far from a creek, ideal for a Mountain Ash.

During its life it has put up with, and beaten, drought, disease, pestilence and the threat of chainsaws. Perhaps most remarkably, it has avoided being burnt in the three brutal bushfires that swept through this area in 1914, 1934 and 1967. Lesser trees would have wilted under the onslaught.

The tree—known to locals as 'the Arve Big Tree'—is 87 metres high, with a circumference at chest height of 17 metres. It is reckoned to have a volume of 368 cubic metres and it probably weighs in excess of 400 tonnes. It is likely this tree is the most massive living thing in Australia. Compare it, however, to the Giant Sequoia (*Sequoiadendron giganteum*) of the United States, the largest of which is a 2500-year-old leviathan nicknamed 'General Sherman' in California's Sequoia National Park. Although only 83 metres tall, it is 31 metres in circumference at the base and has a volume of 1500 cubic metres.

Australia's Mountain Ash are generally regarded as the world's second-tallest trees, after the Coast Redwood (*Sequoia sempervirens*) which live on the west coast of the United States, principally in California and Oregon. The largest known Coast Redwood is 115.6 metres tall so Australia's Mountain Ash—which regularly top

90 metres—is not far behind. In the statistics contest, this tree is the world's tallest hardwood and the tallest flowering plant on Earth.

You can spot a Mountain Ash by its trunk. Attached to the first 15 metres or so of the tree is a rough skirt of bark. Then the trunk turns smooth, and is straight and grey. Branches do not start for about 40 metres.

Eucalyptus regnans are the emperors of Australian forests. They populate the high-rainfall country of Tasmania and Victoria and are usually found in the rich, moist soil of remote mountain valleys. The sight of a stand of Mountain Ash, with their usual understorey of Tree Ferns, remains etched in the memory. And they don't just attack the visual senses. Their bark hangs down in great sheets, up to 30 metres long, swaying and clattering in the wind.

Mountain Ash, known as Swamp Gums in Tasmania, rarely live longer than 450 years, so this tree is nearing the end of its life. Everything about it is inspiring: great strips of bark hang off its trunk, almost to the ground, like a long petticoat. Its surrounds are littered with massive fallen trees, large logs and ferns.

Although this tree is clinging to life, like a crotchety elder, with its top blown off and its branches decimated, it remains both statuesque and awe-inspiring. And like many things in nature of a gargantuan size, it is a world in itself. Hundreds of other plants live off it, including lichens, ferns and mosses, even other trees that sprout from a large burl 3 metres above the ground.

To protect this Tasmanian giant, a public viewing gallery has been erected. We stand, gazing upwards, like courtiers paying homage to an ageing king who may not be with us much longer.

LIKE MANY THINGS IN NATURE OF A GARGANTUAN SIZE, IT IS A WORLD IN ITSELF. HUNDREDS OF OTHER PLANTS LIVE OFF IT, INCLUDING LICHENS, FERNS AND MOSSES, EVEN OTHER TREES THAT SPROUT FROM A LARGE BURL 3 METRES ABOVE THE GROUND.

SPOTTED GUM

Corymbia maculata

Murramarang National Park
New South Wales

THE CIRCUMFERENCE OF THIS 400-year-old Spotted Gum—in the Murramarang National Park south of Termeil on the New South Wales south coast—is, according to ranger David Cunningham, 'fourteen hippies around'.

The tree, nicknamed 'Old Blotchy', became a *cause célèbre* some years ago when, according to some, there was a good chance it would be cut down. Tree huggers mobilised, their efforts eventually leading to the announcement that 290 hectares of bushland—including the tree—would be added to the Murramarang National Park, protecting the land in perpetuity.

This fine tree will now live out its life in this charmed forest 6 kilometres from the Pacific Ocean, surrounded by Blue Gums, Ironbarks, Blackbutts, Turpentines and prehistoric-looking Cabbage Palms.

Despite its fame, it is not easy to find. The local rangers have deliberately not advertised its location, for fear of it being loved to death. Too many visitors can compact the soil around a tree, injuring its roots and dramatically shortening its life.

Spotted Gums, which grow principally along the New South Wales coast and into south-east Queensland and eastern Victoria, are among Australia's most beautiful native trees. They are close relatives of the equally graceful Lemon-scented Gum, for which they are often mistaken. If in doubt which is which, crush a leaf.

The Lemon-scented Gum has a distinct citrus smell. Both were formerly classified as eucalypts until 1995, when they became two of 113 species transferred to the new genus *Corymbia*.

It is something of a miracle that this tree survived the late 1800s. Back then this area was plundered by loggers, the logs transported by bullock dray and railway to the steam-powered mills on the coast. The timber was then taken by boat to Sydney where it was used to build the city's wharves. This tree probably survived because it was too big for the crosscut saws of the day.

By any measure the tree is an imposing one, more than 50 metres high and 9 metres in circumference at the base. Thirty metres of massive straight trunk reaches skywards before the first branches appear. Its bark is smooth and powdery, in shades of cream, blue and white. It is a beautifully complete tree, despite losing two or three big branches over the years, each one weighty enough to be a tree in its own right.

Its trunk is twisted, probably due to the seasons and the sun and, near its base, the bark is wrinkled and pinched, like a waterfall caught on camera. When the bark is injured—perhaps from a falling branch or a love-struck teenager with a penknife—the replacement bark flows over the wound like honey.

Spotted Gums tend to drop limbs on hot, humid days because the moisture in the timber expands and the branches become too heavy for the tree to hold. More's the pity if this ever happened to 'Old Blotchy', whose upper branches would be home to all sorts of marsupials, including possums and sugar gliders.

THIS TREE PROBABLY SURVIVED BECAUSE IT WAS TOO BIG FOR THE CROSSCUT SAWS OF THE DAY.

FLOODED GUM
Eucalyptus grandis

Wang Wauk State Forest
New South Wales

FLOODED GUMS, ALSO KNOWN AS ROSE GUMS, ARE COMMON along New South Wales river flats and along the myriad of escarpments between Newcastle and south-east Queensland. They rarely grow taller than 55 metres, which makes this Flooded Gum, 6 kilometres down Stoney Creek Road north-west of Bulahdelah in the extensive Wang Wauk State Forest, all the more remarkable.

The tree is almost 80 metres tall, which many claim makes it the tallest tree in New South Wales. It is magnificently erect and has a 9-metre circumference at chest height. Experts put the tree's age at more than 400 years.

The tree, which lives within earshot of the dull rumbling of traffic on the Old Pacific Highway, is a living link with the old-growth forest that once existed in this area of New South Wales. It lives in a wet sclerophyll forest, surrounded by giant Bangalow Palms and thick mid-layer shrubs that stop strong light penetrating the forest floor. The damp ground crawls with blood-sucking leeches.

This Flooded Gum probably began its life after a fire came through the forest, or after a giant tree fell down. Such events break open the canopy and let light into the dense forest. It then becomes a race against time. Young trees have to grow quickly to get their crowns into the sunlight before the shrub layer re-forms around them. Consequently Flooded Gums are among the fastest growing of all eucalypts, sometimes growing up to 3 metres a year. Once established, they muscle into the surrounding shrubs by expanding their roots and crowns. Their trunks thicken considerably in order to support the rapid new growth.

The tree's bark is thin and deciduous. Flooded Gums shed their bark each year, in early summer, revealing a smooth powdery trunk with flowing patterns of silver, white, grey and light green. Hollowed-out branches near this tree's crown provide vital roosting and nesting sites for native birds and marsupials such as possums and gliders. Rough fibrous bark covers the lowest 5 metres of the tree's trunk, and strips of recently shed bark hang down from the forks of branches. The timber of the Flooded Gum is rose red. It is softer than many other eucalypt timbers, which makes it easy to work with. Flooded Gum has become an important plantation timber around the world, and huge planting programs have been carried out in South Africa and Brazil. There are also substantial plantings in Angola, Argentina, India, Uruguay, Zaire and Zambia.

FLOODED GUMS ARE AMONG
THE FASTEST GROWING OF ALL
EUCALYPTS, SOMETIMES GROWING
UP TO 3 METRES A YEAR.

DEANE'S BLUE GUM

Eucalyptus deanei

Blue Mountains
New South Wales

IF YOU FIND YOURSELF FLYING above the Blue Mountains west of Sydney, over the town of Blackheath, keep an eye out for a giant Deane's Blue Gum. It is clearly visible sticking out of the bushland in the valley floor to the north-west of the township.

The tree grows in a small amphitheatre at the bottom of a steep escarpment, two hours' walk down an unnamed spur from Perry's Lookdown. It is one of close to 2000 Deane's Blue Gums in the 40-hectare Blue Gum Forest.

About 100 metres from the tree's base is the Grose River, which winds its way east from the flanks of Mount Victoria to the Hawkesbury River. In heavy rains the river floods the amphitheatre in which the trees live. Deane's Blue Gums love both water and rich alluvial soil, so this is ideal country for them.

The tree lives beside the convergence of four walking tracks and is an important landmark for hikers. It is the perfect place for them to shed their hiking packs to sit in the tree's shade and marvel at its scale.

Unwittingly, they might just have saved its life. In 2006 a bushfire raged through the forest, burning the ground cover and much of the forest canopy. But it did not destroy the tree. Local rangers say over the years the

hikers' boots had kept the surrounding undergrowth to a minimum, reducing the impact of the fire.

Deane's Blue Gums, also known as Round-leaved Gums, have a restricted distribution in eastern New South Wales. Not to be confused with Sydney Blue Gums (*Eucalptus saligna*), Deane's Blue Gums typically live in tall wet forests or in sheltered valleys, preferably in deep sandy alluvial soil. They are known for their smooth blue-white bark, and the sock of rough bark at their base. They shed their bark at the beginning of spring.

This tree, 65 metres tall and 6 metres in circumference at chest height, towers above its companions in the forest. Ring counts on smaller trees that have fallen suggest it could be more than 600 years old. Its massive trunk is pockmarked with several fine burls, some more than a metre in diameter.

For more than twenty-five years the dramatic escarpments of the Blue Mountains formed the western boundary of the first settlement of Sydney, its cliffs repelling all who tried to cross them. The region's dramatic landscape has been forged by millions of years of changing climatic conditions including ice ages, searing heat and the collision of continental plates.

THE TREE LIVES BESIDE THE CONVERGENCE OF FOUR WALKING TRACKS AND IS AN IMPORTANT LANDMARK FOR HIKERS. UNWITTINGLY, THEY MIGHT JUST HAVE SAVED ITS LIFE.

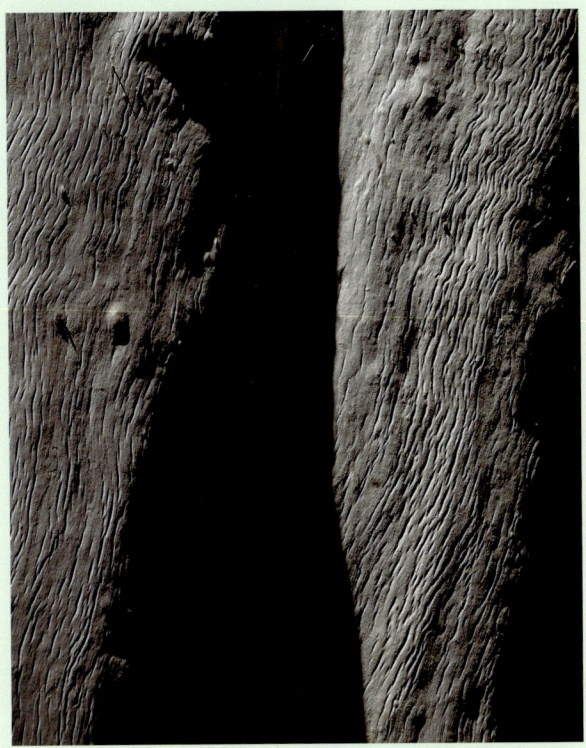

In 1793 William Paterson, a soldier–explorer and enthusiastic botanist, followed the Hawkesbury River from the coast in a whaleboat, exploring and naming the Grose River. In 1813 Gregory Blaxland, William Wentworth and William Lawson set out from near St Marys and took seventeen days to reach Mount York via high forested ridges, opening up much of the area for farming.

Even at this early stage the tree had survived many fires and floods. Since the area was found to be suitable for agricultural purposes and European settlement, the tree has also survived timber cutting (mainly for fence posts), 140 years of cattle grazing and even plans for a Macadamia nut farm.

In November 2000 the Blue Gum Forest probably played a part in persuading UNESCO to put the Greater Blue Mountains on the World Heritage List. The area covers one million hectares, and includes ninety-one species of gum tree. Today, the Deane's Blue Gums in the Blue Mountains are protected, alongside other trees including Angophoras, Ironbarks, Turpentines, banksias and casuarinas.

'The best friend on Earth of man is the tree' —Frank Lloyd Wright